2/∞

Lattice Path Counting
and Applications

This is a volume in
PROBABILITY AND MATHEMATICAL STATISTICS

A Series of Monographs and Textbooks

Editors: Z. W. Birnbaum and E. Lukacs

A complete list of titles in this series appears at the end of this volume.

Lattice Path Counting and Applications

SRI GOPAL MOHANTY
Department of Mathematics
McMaster University
Hamilton, Ontario, Canada

 1979

ACADEMIC PRESS

A Subsidiary of Harcourt Brace Jovanovich, Publishers

New York   London   Toronto   Sydney   San Francisco

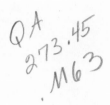

ACADEMIC PRESS, INC.
111 Fifth Avenue, New York, New York 10003

United Kingdom Edition published by
ACADEMIC PRESS, INC. (LONDON) LTD.
24/28 Oval Road, London NW1 7DX

Library of Congress Cataloging in Publication Data

Mohanty, Gopal.
  Lattice path counting and applications.

  (Probability and mathematical statistics)
  Includes bibliographies and index.
  1. Combinatorial probabilities. 2. Lattice paths.
I. Title.
QA273.45.M63   519.2   79–23524
ISBN 0–12–504050–4

PRINTED IN THE UNITED STATES OF AMERICA

79 80 81 82   9 8 7 6 5 4 3 2 1

To my parents

Everything should be made as simple
as possible, but not simpler.

Albert Einstein

Contents

Preface

In the recent past, we witnessed a phenomenal development that indicated the simple nature of a combinatorial approach to a number of diversified problems, mostly in probability theory and statistics. It is not inappropriate to say that the new beginning in this direction was made by the publication of the second edition of W. Feller's book *An Introduction to Probability Theory and Its Applications,* Vol. 1, in which the author added a new chapter (viz., Chapter 3) in order to provide examples of the "newly discovered power of combinatorial methods." In Feller's words (p. 256), it is instructive to realize "that a most elementary combinatorial argument enabled us to solve a difficult technical problem and that it replaces a formidable analytical apparatus."

Subsequent to Feller's book, which is rich in variety as well as in depth, the publication of *Combinatorial Methods in the Theory of Stochastic Processes* by L. Takács amply substantiated Feller's observation. In addition to these books, a large volume of research work has since emerged; this is the result of using the newly discovered power, each time more effectively and diversely than in the past. When we mention combinatorial techniques, we do not refer to the well-known cases of the construction of design of experiments. Rather the new tool may be termed "the ballot theorem," which essentially involves counting lattice paths under certain constraints.

While probabilitists and statisticians were investigating and testing the strength of their newly acquired power, the combinatorialists often independently studied various other aspects of the same topic and have made their share of the contributions. J. Riordan's book *Combinatorial Identities* is an excellent example of this.

With this background, I was initiated into this novel field through T.V. Narayana, University of Alberta, whose ideas and encouragement have influenced me throughout these years. It was the late L. Moser's constant interest in my work while I was a student that led me to the fascinating list of combinatorial identities by H.W. Gould, which played a significant role in my later work. B.R. Handa as my student and collaborator has helped me tremendously to sustain my interest in the field, where on the one hand, one has to keep an eye on the works of combinatorialists and on the other that of probabilitists and statisticians. Frequent discussions and joint works with I.Z. Chorneyko, an esteemed friend and colleague, over the past several years have indeed been a great pleasure. Personal communication and discussion with L. Takács, I. Vincze, E. Sparre Andersen, L. Comtet, G. Kreweras, H.W. Gould, R. Pyke, G.P. Steck, M. Dwass, and N.C. Severo have been a constant inspiration for my present work.

Lattice path models frequently arise in various fields of study, and counting lattice paths under certain restrictions becomes a useful problem in those cases. Apart from the books mentioned above, there has been a growing need to synthesize and summarize the results of the past two decades related to lattice path combinatorics. To my greatest delight, now we see T.V. Narayana's monograph *Lattice Path Combinatorics with Statistical Applications* (University of Toronto Press), in which he has dealt with problems which could be conveniently treated in terms of "domination" (see Chapter 1, Section 6). It is but natural that some of the content of the present work overlaps that of Narayana as well as that of Feller and Takács, which is neither avoidable nor inappropriate.

In Chapter 1 various methods of counting paths with simple boundaries and with diagonal steps are discussed. Chapter 2 is a continuation of Chapter 1, in which we deal with some counting methods for general boundaries, paths with several types of diagonal steps, and continuous analogues of path counting. Certain invariance properties of paths and their relations to fluctuation theory are the subject matter of Chapter 3. Various fields of application, viz., random walk, rank order statistics, discrete distributions, queues, trees, and search codes, are dealt with in Chapters 4 and 5. In Chapter 6 we discuss and develop certain identities and inverse relations from path combinatorics. Each chapter is appended with a set of exercises and a list of references. In a way, a humble attempt is made to tie together the loose ends in lattice path combinatorics. In the process, some new loose ends could possibly have crept in.

Acknowledgments

My sincere appreciation is expressed to the Indian Statistical Institute, the Indian Institute of Technology, Delhi, the University of Delhi, the Mathematical Institute of the University of Copenhagen, and the Mathematical Institute of the Hungarian Academy of Sciences. During my sabbatical leave (1974–1975), the major part of the writing was done at these places. Parts of the material were presented in a series of seminars at the Indian Statistical Institute and the University of Delhi. Special mention is made of J.K. Ghosh and A.R. Rao of the Institute and H.C. Gupta, Kanwar Sen, and J.L. Jain of the University,whose interests in the seminars were very encouraging. Finally, I owe my deep gratitude to Z.W. Birnbaum, whose personal interest helped the publication of this monograph.

I gratefully acknowledge the financial contributions of the National Research Council, Canada, over several years, which has enabled me to continuously engage in research in this field and complete this piece of work.

It is a pleasure to record my personal appreciation to Mrs. J. Fabricius and Miss J. Maljar who have typed the final manuscript.

1 Path Counting—Simple Boundaries

1. Introduction

Lattice paths are encountered in a natural way in various problems, e.g., ballot problems, compositions, random walks, fluctuations, queues. The well-known classical ballot problem of Bertrand [4] can be stated as follows: In an election, two candidates A and B have m and n ($m > n$) votes, respectively; it is required to determine the probability that at each stage of counting A's votes exceed those of B. If A's vote is represented by a horizontal unit and B's by a vertical unit in a plane, the solution of the problem depends on the number of lattice paths from $(0, 0)$ to (m, n) which do not touch the line $x = y$. Also, one can show easily that any such path without any restriction represents a composition of the integer $m + n + 1$ into n parts.

Lattice paths under certain restrictions appear in many situations in probability and statistics. Sometimes it might be necessary to transform a given problem or part of it into lattice paths. In this chapter we deal only with enumeration of lattice paths within certain boundaries and their relation to and interpretation as ballot problems, compositions of an integer, and Fibonacci numbers. As we proceed further, other applications will be developed through discussions, results, and exercises.

1

2. Two-Dimensional Lattice Paths

In the two-dimensional space, by a lattice path (briefly, a path) from the lattice point (x_1, y_1) to the lattice point (x_2, y_2), $x_1 \leq x_2, y_1 \leq y_2$, we mean a directed path from (x_1, y_1) to (x_2, y_2) which passes through lattice points with movements parallel to the positive direction of either axis. Here, we may refer to two types of steps, viz., x-steps and y-steps, where an x (y)-step is a line segment parallel to the x (y) axis joining two neighboring points.

For counting purposes we may, without loss of generality, consider paths from the origin to (m, n) and observe that each such path is characterized by having exactly m x-steps and n y-steps. If we denote by $L(m, n)$ the set of paths from $(0, 0)$ to (m, n) and by $|S|$ the number of elements in S, elementary reasoning gives the result

$$|L(m, n)| = \binom{m + n}{n}. \tag{1.1}$$

3. Reflection Principle (Conjugation)

Let us count paths from the origin to (m, n), $m > n + t$, not touching the line $x = y + t$, where t is a nonzero integer. Denote this set of paths by $L(m, n: t)$. The number is obtained simply by subtracting from the total number of paths from $(0, 0)$ to (m, n) the number of paths that reach the line $x = y + t$. Denote the set of latter paths by $R(m, n; t)$. To evaluate $|R(m, n; t)|$ we use a mapping induced by the so-called *reflection principle* introduced by André [2] for solving the original ballot problem.

Referring to Fig. 1 (t is taken to be negative), one can observe that every path in $R(m, n; t)$ must reach the line $x = y + t$ for the first time, say at $(j, j - t), j = 0, 1, \ldots, n + t$, before reaching (m, n). Consider a given path which reaches the boundary for the first time at $(j, j - t)$. The path consists of two parts: (i) from $(0, 0)$ to $(j, j - t)$ and (ii) from $(j, j - t)$ to (m, n). The mapping of this path to a new path is as follows:

Change (i) to its reflection about the line $x = y + t$ so that the lattice point (u, v), $0 \leq u \leq j$, $0 \leq v \leq j - t$ on the path becomes $(v + t, u - t)$;

Retain (ii) as it is.

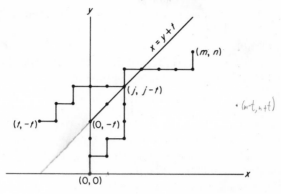

Figure 1

It is easy to verify that the new path is a path from $(t, -t)$ to (m, n) and the mapping is $1:1$ (i.e., bijective). This fact in short can be written as $R(m, n; t) \Leftrightarrow L(m - t, n + t)$ and thus

$$|R(m, n; t)| = |L(m - t, n + t)| = \binom{m + n}{m - t} \qquad (1.2)$$

which is also valid when $t > 0$. It leads us to the next result.

THEOREM 1

$$|L(m, n: t)| = \binom{m + n}{n} - \binom{m + n}{m - t}, \qquad t \neq 0. \qquad (1.3)$$

When $t = 0$, the paths have to touch the line $x = y$ at the origin, and therefore the number of paths from the origin to (m, n) that do not touch the line $x = y$ except at the origin is given by

$$|L(m - 1, n: -1)| = \frac{m - n}{m + n}\binom{m + n}{n}. \qquad (1.4)$$

The reflection acts like an operator on a path that changes x-steps to y-steps and y-steps to x-steps. For a given path p, path p^* which is constructed by interchanging x-steps and y-steps is said to be the conjugate of p (p^* is the reflection of p about $x = y$). When the line about which the reflection is taken is ignored, we may use the word *conjugation* in place of reflection.

In Section 1 we indicated the connection between the lattice paths and the ballot problem. Thus the probability that at each stage of counting A's votes exceed B's votes is given by $(m - n)/(m + n)$. For historical notes on ballot problems one may see Feller [10] and Takács [41].

It is possible to derive the number of paths that do not cross the boundary from $|L(m, n: t)|$. The number of paths from $(0, 0)$ to (m, n) not crossing the line $x = y$ can be checked to be

$$\frac{m - n + 1}{m + n + 1} \binom{m + n + 1}{n}. \tag{1.5}$$

Whitworth's book [44] contains (1.5). We may remark that in two-dimensional paths, the boundary not to be crossed is equivalent to a corresponding boundary not be touched.

An important special case arises when $m = n$ and the number of paths is $(1/(n + 1))\binom{2n}{n}$, which for various values of n are known as Catalan numbers. These numbers come across in many combinational problems (see Alter [1]). For example, Catalan's original problem was to determine the number of ways that the sequence $x_1, x_2, \ldots, x_n$ can be combined in this order by a binary nonassociative product. The answer $(1/n)\binom{2n-2}{n-1}$ is a Catalan number. For $n = 4$, the possible cases are:

$$(x_1(x_2(x_3 x_4))), \qquad (x_1((x_2 x_3)x_4)), \qquad ((x_1(x_2 x_3))x_4),$$
$$(((x_1 x_2)x_3)x_4), \qquad ((x_1 x_2)(x_3 x_4)).$$

According to the formula, the number is $\frac{1}{4}\binom{6}{3} = 5$.

The reader is referred to Brown [6], for references to earlier works and some historical details. Some generalizations of these numbers are possible in various directions (see [7, 8]).

It is interesting to remark that there is a $1:1$ correspondence between a standard Young tableau of shape (m, n) and a path from $(0, 0)$ to (m, n) that does not cross the line $x = y$. A standard Young tableau of shape $(\alpha_1, \ldots, \alpha_r)$, $\alpha_1 \geq \alpha_2 \geq \cdots \geq \alpha_r$, $\sum_{i=1}^{r} \alpha_i = k$, is an arrangement of $1, 2, \ldots, k$ in r rows with the ith row having α_i cells in the manner shown in the diagram, such that the cell entries in rows and columns are in increasing order.

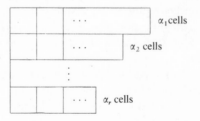

For example, a Young tableau of shape $(3, 2, 2)$ follows. (We have omitted "standard" for brevity.)

1	2	4
3	6	
5	7	

To show the correspondence one uses the following construction. The cells are filled in from left to right in each row. If the ith step in a given path is an x-step, enter i in the first unfilled cell in the first row. On the other hand if it is a y-step, enter i in the first unfilled cell in the second row. The construction ensures 1 : 1 correspondence. An illustration now follows.

Path from (0.0) to (5.3)   Young tableau of shape (5.3)

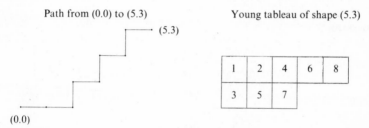

(5.3)

1	2	4	6	8
3	5	7		

(0.0)

In a Young tableau the hook length of the (i, j)th cell, denoted by h_{ij}, is defined to be

$h_{ij} = 1 +$ number of cells to the right of the (i, j)th cell
$+$ number of cells below the (i, j)th cell.

An important combinatorial theorem due to Frame *et al.* [11] states that the number of Young tableaux of shape $(\alpha_1, \alpha_2, \ldots, \alpha_r)$, where $\sum_{i=1}^{r} \alpha_i = k$ is equal to

$$\frac{k!}{\prod_i \prod_j h_{ij}}. \tag{1.6}$$

The table of hook lengths corresponding to paths is

$m + 1$	m	$\cdots$	$m - n + 3$	$m - n + 2$	$m - n$	$\cdots$	1
n	$n - 1$	$\cdots$	2	1			

Using the above theorem, the number of paths from $(0, 0)$ to (m, n) not crossing the line $x = y$ is given by

$$\frac{(m + n)!}{\dfrac{(m + 1)!}{m - n + 1} n!} = \frac{m - n + 1}{m + n + 1} \binom{m + n + 1}{n}.$$

Result (1.6) is obtained in a different form in [26, 46]. See Barton and Mallows [3] and Kreweras [24] for more information on Young tableaux. In fact, the paper [3] contains a survey of several results interconnected through combinatorics.

By using the reflection principle (or conjugation) repeatedly and the inclusion–exclusion method for counting, we can evaluate the number of paths between two boundaries of the above type. For this purpose denote by $L(m, n; t, s)$ the set of paths from the origin to (m, n) that do not touch the lines $x = y + t$ and $x = y - s$, $t > 0$, $s > 0$.

THEOREM 2

$$|L(m, n; t, s)| = \sum_k \left[\binom{m + n}{m - k(t + s)}_+ - \binom{m + n}{n + k(t + s) + t}_+ \right], \quad (1.7)$$

where

$$\binom{y}{z}_+ = \begin{cases} \dbinom{y}{z} & \text{when } y \geq z, \\ 0 & \text{when } y < 0 \ \text{ or } \ y < z, \\ 1 & \text{when } z = 0, \end{cases}$$

and the summation is over all integer values of k: positive, negative, and zero.

Proof

For brevity, call the boundaries $x = y + t$ and $x = y - s$, L^+ and L^-, respectively. Denote by A_1 the set of paths that reach L^+, by A_2 the set of paths that reach L^+, L^- in that order, and in general by A_i the set of paths reaching $L^+, L^-, L^+, \ldots$ (i times) in the specified order. Similarly, let B_i be the set of paths reaching $L^-, L^+, L^-, \ldots$ (i times) in the specified order. An application of the usual inclusion–exclusion method yields

$$|L(m, n; t, s)| = \binom{m + n}{n} + \sum_{i \geq 1} (-1)^i (|A_i| + |B_i|), \quad (1.8)$$

where $|A_i|$ and $|B_i|$ are evaluated by using the reflection principle repeatedly. For example, consider A_3. Since every path in A_3 must reach L^+, A_3 when reflected about L^+ becomes the set of paths from $(t, -t)$ to (m, n) each of which reaches L^+ after reaching L^-. Another reflection about L^- would make A_3 equivalent to the set of paths from $(-s - t, s + t)$ to (m, n) that reach L^+, which in turn can be written as $R(m + s + t, n - s - t; 2s + 3t)$. Thus by (1.2),

$$|A_3| = \binom{m + n}{m - s - 2t}$$

and in general

$$|A_{2j}| = \binom{m + n}{m + j(s + t)}_+, \qquad |A_{2j+1}| = \binom{m + n}{m - j(s + t) - t}_+. \qquad (1.9)$$

The expressions for $|B_{2j}|$, $|B_{2j+1}|$, $j = 0, 1, 2, \ldots$, with $|A_0|$, $|B_0|$ being $\binom{m+n}{n}$, are obtained by interchanging m with n and s with t. Substitution of these values in (1.8) yields (1.7) after some simplifications. This completes the proof. Also see [40].

The mapping induced by the reflection principle may not be bijective for some boundaries, and in such cases we should look for alternative methods of counting.

4. Method of Penetrating Analysis

Consider the set of paths from the origin to (m, n), $m > \mu n$, not touching the line $x = \mu y$, μ being a nonnegative integer, and let the set be denoted by $L(m, n; \mu)$. It may be checked that there is no easy way to use the reflection principle of the previous section in order to enumerate the paths that touch or cross the line $x = \mu y$. An alternative approach, which was suggested by Dvoretzky and Motzkin [9], is called (as Grossman [20] puts it) *penetrating analysis* and is described by an illustration. Let us take the example with $m = 9$, $n = 4$, and $\mu = 2$ for which a typical path is PQ in Fig. 2. QR is a repetition of PQ. Any path starting from a point in PQ excluding Q and ending after 13 units is said to be a cyclic permutation of the path PQ. Thus there are 13 ($=9 + 4$) cyclic permutations of PQ.

Imagine an infinitely distant light source in the positive direction of the line $x = 2y$ (in general $x = \mu y$) that is hitting PQR, which is supposed to be an opaque screen. Then all points in PQ except D are in

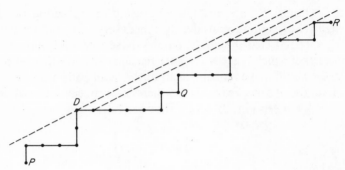

Figure 2

shade. A little reflection will show that four vertical units block eight horizontal units, leaving only one in the light. Clearly among all 13 cyclic permutations of PQ the only path beginning at D will lie below the line $x = 2y$.

From the above analysis it is also evident that even if we would have started with any path other than PQ, among its 13 cyclic permutations, there would be only one which would lie below the line $x = 2y$.

In general, we divide the $(m + n)!$ paths that arise when various x-steps and y-steps are permuted (treating each as distinct) into $(m + n - 1)!$ sets of $m + n$ paths that are cyclic permutations of each other. In each set only $m - \mu n$ paths will not touch the line $x = \mu y$. Thus the probability of a path from the origin to (m, n) that does not touch the line $x = \mu y$ is $(m - \mu n)/(m + n)$. Hence the next result in path counting is as follows:

THEOREM 3

$$|L(m, n; \mu)| = \frac{m - \mu n}{m + n} \binom{m + n}{n}. \qquad (1.10)$$

In terms of ballot problems the theorem states that the probability for A's votes to exceed μ times B's votes is $(m - \mu n)/(m + n)$.

As early as 1941 (Lyness [25]) this result appeared in an uncommon but imaginative form. Al Capone sets out by armored car from $(0, 0)$ for the rival gang's headquarters at $(n, \mu n + c)$. Unknown to him, the rival gang has a death ray sent out from its headquarters in the direction of (arccot μ). The chance of Al's getting to his enemy alive is determined by (1.10), with $c = m - \mu n$.

By the duality principle (as in Feller [10, p. 70]), almost every result on paths can be reformulated to obtain a formally different one. Let us consider a dual counting problem. For a given path p, path p' which is constructed from p by taking steps in reverse order is said to be the reverse of p. Then the set of paths, the reverse of those in Theorem 3, are the paths from $(0, 0)$ to (m, n) not touching the line $x = \mu y + k$ ($k = m - \mu n$) except at the end. Denote this set by $L'(k, n; \mu)$. Then Theorem 3 is equivalent to

$$|L'(k, n; \mu)| = \frac{k}{k + (\mu + 1)n} \binom{k + (\mu + 1)n}{n}. \tag{1.11}$$

We come back to this dual problem in the next section. Also we note that if we consider conjugate paths, we have formally a new result. Thus duality may consist of either reversion or conjugation.

Note that the method of penetrating analysis has the following features:

(a) the application of a combinatorial trick;
(b) the invariance property for each set of cyclically permutated paths;
(c) the use of a probabilistic approach in order to obtain a combinatorial result.

Later, the preceding demonstrative argument will be refined and presented more rigorously for a general situation.

In (1.11) we have obtained an expression for the number of paths from the origin to a lattice point on the line $x - \mu y = k$ ($\mu \geq 0$), not touching the line except at the end. Consider μ to be a negative integer. Let $f_\mu(k, y)$ be the number of paths from $(0, 0)$ to $(k + \mu y, y)$, which is on the boundary $x - \mu y = k$. Taking $\mu = -r$ (r being a positive integer), one can easily establish

$$\begin{aligned}
f_r(k, y) &= 0, \quad k < 0, \quad y < 0, \\
f_r(0, 0) &= 1, \\
f_r(k, y) &= f_r(k - 1, y) + f_r(k - r, y - 1),
\end{aligned} \tag{1.12}$$

and

$$\begin{aligned}
f_r(k) &= 0, \quad k < 0, \\
f_r(0) &= 1, \\
f_r(k) &= f_r(k - 1) + f_r(k - r),
\end{aligned} \tag{1.13}$$

where

$$f_r(k) = \sum_y f_r(k, y).$$

For $r = 1$ (1.12) generates the well-known Pascal triangle. For $r = 2$ one can see from (1.13) that $\{f_2(k)\}$ is the sequence of Fibonacci numbers, and therefore $\{f_2(k, y)\}$, for fixed k, forms a partition of Fibonacci number $f_2(k)$. Thus for general r we may say that $f_r(k)$ is a generalized Fibonacci number. A finer partition of a generalized Fibonacci number is given in [28].

5. Recurrence and Generating Function Method

Generally, any counting problem is solved either by a set of recurrence relations or by the generating function technique or both. In our situation if we denote by $\alpha(m, n | \cdot)$ the set of lattice paths from the origin to (m, n), restricted by the condition $(\cdot)$, the following recurrence relation is easily established:

$$|\alpha(m, n | \cdot)| = |\alpha(m, n - 1 | \cdot)| + |\alpha(m - 1, n | \cdot)| \qquad (1.14)$$

with corresponding boundary conditions. Using (1.14), one can, for example, derive (1.10) simply by induction. However, it is difficult to guess the result in advance so that an inductive proof could work. Even though explicit solution of (1.14) is not readily obtainable in all situations, the relation is often useful for computation with the aid of a modern computer.

Regarding the applicability of the generating function, let us again consider $|L'(k, n; \mu)|$ which for simplicity in notation is written as $(k; n)$. $(k; n)$ satisfies, for $k = 1, 2, \ldots$,

$$\begin{aligned}
(k; 0) &= (k - 1; 0) = 1, \\
(k; n) &= (k - 1; n) + (k + \mu; n - 1), \qquad n = 1, 2, \ldots.
\end{aligned} \qquad (1.15)$$

Instead of the usual bivariate generating function on k and n, we take the generating function on the number of steps in a path, the reason for which will be clear in the next paragraph. Thus multiplication of $s^{(\mu + 1)n + k}$ on both sides of (1.15) and summation over n yield

$$G(s; k) = sG(s; k - 1) + sG(s; k + \mu), \qquad (1.16)$$

where

$$G(s; k) = \sum_{n=0}^{\infty} (k; n)s^{(\mu + 1)n + k}.$$

An important relation on $(k; n)$ is (i.e., convolution identity)

$$(k; n) = \sum_{r=0}^{n} (a; r)(k - a; n - r), \qquad (1.17)$$

a being an integer such that $0 < a < k$. Therefore

$$G(s; k) = G(s; a)G(s; k - a)$$
$$= [G(s; 1)]^k, \qquad (1.18)$$

the last step being obtained by repeated use of the first part. Using (1.18), relation (1.16) is reduced to

$$s[G(s; 1)]^{\mu+1} - [G(s; 1)] + s = 0, \qquad (1.19)$$

the solution of which will be obtained by the Lagrange series expansion (see Pólya and Szegö [35, p. 125]).

The Lagrange series expansion of a function $f(x)$ is

$$f(x) = f(0) + \sum_{n=0}^{\infty} \frac{y^n}{n!} \left[\frac{d^{n-1}}{dx^{n-1}} (f'(x)\phi^n(x)) \right]_{x=0}, \qquad (1.20)$$

where $y = x/\phi(x)$ (see Riordan [36, p. 146]). For a mathematical treatment the reader may refer to Whittaker and Watson's book [43, pp. 132–133].

Relation (1.19) can be written as

$$s^{\mu+1} = \left[\frac{G(s; 1)}{s} - 1 \right] \Big/ \left(\frac{G(s; 1)}{s} \right)^{\mu+1}.$$

By putting $y = s^{\mu+1}$, $x = [G(s; 1)/s] - 1$,

$$\phi(x) = \left(\frac{G(s; 1)}{s} \right)^{\mu+1} = (1 + x)^{\mu+1}, \qquad f(x) = (1 + x)^k,$$

and applying (1.20), we get

$$\left(\frac{G(s; 1)}{s} \right)^k = 1 + \sum_{n=1}^{\infty} s^{n(\mu+1)} \frac{k}{n} \binom{n(\mu + 1) + k - 1}{n - 1},$$

which simplifies to

$$G(s; k) = \sum_{n=0}^{\infty} \frac{k}{k + (\mu + 1)n} \binom{k + (\mu + 1)n}{n} s^{k+(\mu+1)n}. \qquad (1.21)$$

Thus

$$|L'(k, n; \mu)| = \frac{k}{k + (\mu + 1)n} \binom{k + (\mu + 1)n}{n},$$

which was derived before by another method.

We remark that the generating function technique may not be successfully applied to any general boundary situation. It depends on writing an equation of type (1.19) and finding an explicit solution for the desired generating function, which may not always be simple. When $\mu = 1$, (1.19) is a quadratic equation and is easily solvable (see [10, p. 319]).

It is instructive to give Bizley's result [5], as an illustration of a combination of various techniques (say, penetrating analysis and generating function) in solving a problem. Suppose we are interested in the number of paths (denoted by ψ_k) from $(0, 0)$ to (km, kn) that do not touch the line $my = nx$, except at the endpoints, where k is a positive integer and m and n are coprime positive integers. Let us denote by $\phi_{k,t}$ the number of paths that touch the line $my = nx$ at t points, including the last one, but do not cross the line. Any such path is called a path with t contacts. In this notation $\psi_k = \phi_{k,1}$. We say that (x_0, y_0) is a highest point in the path, if $y_0 = (n/m)x_0 + r$ for some r and $y \leq (n/m)x + r$ for every point (x, y) on the path.

Under the penetrating analysis, we note the following:

(i)   the number of highest points on a path remains invariant by a cyclic permutation;

(ii) any path with t highest points can be transformed into a path with t contacts by a cyclic permutation in exactly t ways. (This is done when a cyclic permutation starts with a highest point as the origin.)

When we consider $k(m + n)\phi_{k,t}$ paths by permuting cyclically all $\phi_{k,t}$ paths, we observe that every path with t highest points will be formed in exactly t times. Hence the number of paths with t highest points is $k(m + n)\phi_{k,t}/t$, which implies

$$\binom{km + kn}{km} = \sum_{t=1}^{k} \frac{k(m + n)}{km} \phi_{k,t}, \qquad (1.22)$$

the left-hand side being the number of paths from $(0, 0)$ to (km, kn). Letting

$$F_k = \frac{1}{k(m + n)} \binom{km + kn}{km},$$

(1.22) becomes

$$F_k = \sum_{t=1}^{k} \frac{1}{t} \phi_{k,t}. \qquad (1.23)$$

Also, we have

$$\phi_{k,t} = \sum \psi_{a_1}\psi_{a_2}\cdots\psi_{a_t}, \tag{1.24}$$

the sum extending over $(a_1,\ldots,a_t)$ such that $a_i > 0$, $i = 1,\ldots,t$, and $\sum_{i=1}^{t} a_i = k$.

Let $H(s)$ be the generating function of $\{\psi_k\}$, i.e.,

$$H(s) = \sum_{j=1}^{\infty} \psi_j s^j, \tag{1.25}$$

and we want to find an explicit expression for $H(s)$. From (1.24) and (1.25) it follows that $\phi_{k,t}$ is the coefficient of s^k in $(H(s))^t$, and therefore the right-hand side of (1.23) is the coefficient of s^k in $\sum_{t=1}^{k} (H(s))^t/k$ or in

$$\sum_{t=1}^{\infty} \frac{1}{t}(H(s))^t = -\log(1 - H(s)).$$

In other words,

$$-\log(1 - H(s)) = \sum_{j=1}^{\infty} F_j s^j, \tag{1.26}$$

i.e.,

$$H(s) = 1 - \exp\left(-\sum_{j=1}^{\infty} F_j s^j\right).$$

Thus ψ_k, which is the coefficient of s^k in the expression of $H(s)$, is given by

$$\psi_k = \sum (-1)^{1 + \Sigma k_i} \prod_i \frac{F_i^{k_i}}{k_i!}, \tag{1.27}$$

the sum being over all k_i such that $k_i \geq 0$ and $\sum i k_i = k$. As a simple corollary of (1.27), we obtain the number of paths from the origin to (m, n) (m and n being coprime) such that no paths touch the line $nx = my$ as $(m + n)^{-1}\binom{m+n}{n}$.

If we differentiate both sides of (1.26) with respect to s, multiply through by $(1 - H(s))$, and equate coefficients of successive powers of s, we obtain the recurrence relations on the ψs, the first three of which are

$$(m + n)\psi_1 = \binom{m + n}{n},$$

$$2(m + n)\psi_2 = \binom{2m + 2n}{2m} - \binom{m + n}{m}\psi_1,$$

$$3(m + n)\psi_3 = \binom{3m + 3n}{3m} - \binom{2m + 2n}{2m}\psi_1 - \binom{m + n}{m}\psi_2.$$

One may give a combinatorial proof of the general recurrence formula. Similar expressions can be derived for $\phi_{k,t}$ and are left as exercises.

The technique of vector representation discussed in the next section has a greater appeal for its simplicity as well as for its generality.

6. Vector Representation, Compositions, and Domination

We rederive $|L(m, n; \mu)|$ by yet another approach, which is based on vector representations of paths. For this purpose, associate with a path the vector $(x_1, x_2, \ldots, x_n)$, where x_i $(i = 1, \ldots, n)$ is the minimal distance, measured parallel to the x axis, of the points $(m, n - i)$ from the path. In Fig. 3 a path from $(0, 0)$ to $(8, 5)$ is drawn for which $x_1 = 2$, $x_2 = 3, x_3 = 3, x_4 = 3, x_5 = 6$. Let $S(m, n; \mu)$ be the set $(x_1, x_2, \ldots, x_n)$ of vectors of n elements such that

(i) x_i is an integer, $i = 1, \ldots, n$;
(ii) $0 \le x_1 \le x_2 \le \cdots \le x_n$;
(iii) $x_i \le m - \mu n - 1 + (i - 1)\mu, i = 1, \ldots, n$.

Because the mapping $L(m, n; \mu)$ to $S(m, n; \mu)$ induced by the vector representation is seen to be bijective,

$$|L(m, n; \mu)| = |S(m, n; \mu)|.$$

Before evaluating $S(m, n; \mu)$, we introduce the concept of domination, as developed in [31], on compositions of an integer and its relation with $S(m, n; \mu)$. Given an integer N, an r-composition of N denoted by

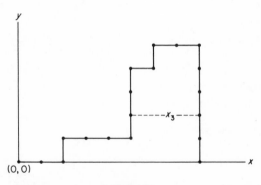

Figure 3

$(t_1, t_2, \ldots, t_r)$ is a vector of r elements, where $t_i \geq 1$ is an integer for $i = 1, \ldots, r$ such that $t_1 + \cdots + t_r = N$. If r is such that $1 \leq r \leq N$, there are obviously $\binom{N-1}{r-1}$ r-compositions of n.

DEFINITION 1

An r-composition $(t_1, t_2, \ldots, t_r)$ of N *dominates* an r-composition $(t'_1, t'_2, \ldots, t'_r)$ of N if and only if

$$\sum_{i=1}^{j} t_i \geq \sum_{i=1}^{j} t'_i, \qquad \text{for} \quad j = 1, \ldots, r. \tag{1.28}$$

Evidently, $t_1 + \cdots + t_r = t'_1 + \cdots + t'_r = N$.

The relation of domination as defined above is reflexive, transitive, and antisymmetric and thus induces a partial order over the set of r-compositions of N. It can also be proved [33] that the set of r-compositions of N forms a distributive lattice ($1 \leq r \leq N$).

As suggested by (1.28), let

$$T_j = \sum_{i=1}^{j} t_i, \qquad j = 1, \ldots, r.$$

Notice that

$$0 < T_1 < \cdots < T_r = N \tag{1.29}$$

and that any set $(T_1, T_2, \ldots, T_r)$ of positive integers satisfying (1.29) corresponds to some r-composition of N. Call $(T_1, T_2, \ldots, T_r)$ the r-composition vector of N corresponding to $(t_1, t_2, \ldots, t_r)$. Clearly, the set of r-compositions and the set of r-composition vectors are in $1 : 1$ correspondence.

Suppose we are interested in counting r-compositions of N which are dominated by a given r-composition $(t_1, t_2, \ldots, t_r)$ of N whose r-composition vector is $(T_1, T_2, \ldots, T_r)$. An important lemma. giving the above number recursively, follows immediately. Denote by D_k, $k \leq r - 1$, the number of vectors $(T'_1, T'_2, \ldots, T'_k)$ such that

   (i) T'_i is a positive integer, $i = 1, \ldots, k$;
   (ii) $T'_1 < T'_2 < \cdots < T'_k$;
   (iii) $T'_i \leq T_i, i = 1, \ldots, k$.

In this notation the required number of r-compositions of N is D_{r-1}.

LEMMA 1

$$D_k = \sum_{i=0}^{k-1} (-1)^i \binom{T_{k-i} + i}{i+1}_+ D_{k-i-1}, \qquad k = 1, 2, \ldots, r-1, \quad (1.30)$$

where $D_0 = 1$.

Proof

Consider the following sets:

$$S_j = \{(T'_1, \ldots, T'_k) : T'_1 < \cdots < T'_{k-j}, T'_{k-j+1} \geq \cdots \geq T'_k, T'_i \leq T_i$$
$$\text{for } i = 1, 2, \ldots, k\},$$

$j = 0, 2, \ldots, k$. By the method of inclusion–exclusion, we have

$$|S_0| = |S_1| - |S_2| + \cdots + (-1)^{k-1}|S_k|,$$

where

$$|S_0| = D_k$$

and

$$|S_j| = |\{(T'_1, \ldots, T'_{k-j}) : T'_1 < \cdots < T'_{k-j}, T'_i \leq T_i$$
$$\text{for } i = 1, \ldots, k-j\}|$$
$$\cdot |\{(T'_{k-j+1}, \ldots, T'_k) : T'_k \leq T'_{k-1} \leq \cdots \leq T'_{k-j+1} \leq T_{k-j+1}\}|$$
$$= D_{k-j} \sum_{T'_{k-j+1}=1}^{T_{k-j+1}} \cdots \sum_{T'_{k-1}=1}^{T'_{k-2}} \sum_{T'_k=1}^{T'_{k-1}} 1$$
$$= D_{k-j} \binom{T_{k-j+1} + j - 1}{j}_+.$$

This completes the proof.

Some elementary simplification of (1.30) gives an alternative expression for D_k as

$$D_k = \sum_{i=0}^{k-1} (-1)^i \binom{T_{k-i} - k + i + 1}{i+1}_+ D_{k-i-1}, \qquad k = 1, 2, \ldots, r-1.$$

$$(1.31)$$

As an application of Lemma 1, the following theorem is presented [29].

THEOREM 4

The number of $(n + 1)$-compositions of N dominated by the $(n + 1)$-composition $(a, b, b, \ldots, b, N - a - (n - 1)b)$ of N is

$$\frac{a}{a + nb}\binom{a + nb}{n}. \tag{1.32}$$

Proof

The result is trivially true for $n = 0$, being equal to 1. The number of 2-compositions of N dominated by $(a, N - a)$ is a. Thus, the theorem is true for $n = 1$. Using induction and applying the recursive formula of Lemma 1, we obtain the required number as

$$\binom{a + (n - 1)b}{1}\frac{a}{a + (n - 1)b}\binom{a + (n - 1)b}{n - 1} - \binom{a + (n - 2)b + 1}{2}$$

$$\times \frac{a}{a + (n - 2)b}\binom{a + (n - 2)b}{n - 2} + \cdots + (-1)^{n - 1}\binom{a + n - 1}{n}$$

$$= \sum_{s = 1}^{n}(-1)^{s - 1}\binom{a + (n - s)b + s - 1}{s}\frac{a}{a + (n - s)b}$$

$$\times \binom{a + (n - s)b}{n - s}$$

$$= \frac{a}{n}\sum_{s = 0}^{n}(-1)^{s - 1}\binom{a + (n - s)b + s - 1}{n - 1}\binom{n}{s} + \frac{a}{a + nb}\binom{a + nb}{n}.$$

Collecting the coefficients of $x^{a + n(b - 1)}$ in the expansion of $(1 - x^{b - 1})^n$ $(1 - x)^{-n}$, we can show that the first summation is zero and hence the proof of the theorem.

Now we suggest an elementary mapping, showing the correspondence between the vectors and compositions. Consider the set of $(n + 1)$-compositions of $N \geq m - \mu + n$ dominated by the $(n + 1)$-composition $(m - \mu n, \mu + 1, \mu + 1, \ldots, \mu + 1, N - m + \mu - n + 1)$ of N, and denote this set by $C(m, n; \mu)$. For $(x_1, \ldots, x_n) \in S(m, n; \mu)$ let ζ be the transformation such that

$$\zeta(x_1, \ldots, x_n) = (t_1, \ldots, t_{n+1}),$$

where

$$t_1 = x_1 + 1, \quad t_i = x_i - x_{i-1} + 1, \quad i = 2, \ldots, n,$$

and

$$t_{n+1} = N - \sum_{i=1}^{n} t_i.$$

The next lemma states the desired correspondence, the proof of which is left to the reader.

LEMMA 2

ζ is a bijective mapping from $S(m, n; \mu)$ to $C(m, n; \mu)$.

Thus the expression for $|L(m, n; \mu)|$ is rederived by the use of Lemma 2 and Theorem 4. Interestingly enough, the same result can still be obtained by using a generalized definition of domination, which is done in an exercise in Chapter 2.

Because of Lemma 2, we may define the relation of domination on the paths. For brevity and without any confusion a path whose vector representation is $(x_1, \ldots, x_n)$ is written as a path $(x_1, \ldots, x_n)$.

DEFINITION 2

A path $(x_1, \ldots, x_n)$ *dominates* the path $(y_1, \ldots, y_n)$ if and only if $y_i \leq x_i$ for all i.

Since the Ts in the composition vector $(T_1, \ldots, T_n, N)$ of the composition $(a, b, b, \ldots, b, N - a - (n - 1)b)$ form an arithmetic progression (i.e., A.P.), we may conveniently say the result in Theorem 4 as referred to the one-A.P. case. A result corresponding to the two-A.P. case [29] is as follows:

THEOREM 5

The number of $(p + q + 1)$-compositions of N dominated by the $(p + q + 1)$-composition

$$(a, b, \ldots, b, c, d, \ldots, d, N - a - (p - 1)b - c - (q - 1)d)$$

of N is given by

$D_{p,q}(a, b; c, d)$

$$= \sum_{k=0}^{q} (-1)^k \frac{a}{a + (p + q - 1)b} \binom{a + (p + q - k)b}{p + q - k}$$

$$\times \frac{(q - k + 1)b - c - (q - k)d}{(q - k + 1)b - c - qd + k} \binom{(q - k + 1)b - c - qd + k}{k}.$$

$$(1.33)$$

The proof is left to the reader as an exercise. A general result for the
r-A.P. case can be obtained following the same line of reasoning as in the
two-A.P. case. Adopting the construction procedure of Lemma 2, we
see that $D_{p,q}(a + 1, b + 1; c - a - (p - 1)b + 1, d + 1)$ is the number
of vectors of type $(x_1, \ldots, x_{p+q})$ such that

(i) $0 \leq a_1 \leq \cdots \leq a_{p+q}$;
(ii) $0 \leq a_\alpha \leq a + (\alpha - 1)b$ for $\alpha = 1, 2, \ldots, p; a, b \geq 0$;
(iii) $0 \leq a_\alpha \leq c + (\alpha - p - 1)d$ for $\alpha = p + 1, \ldots, p + q$,
      $c \geq a + (p - 1)b, d \geq 0$.

For simplicity, we may denote this number by $N_{p,q}(a, b, c, d)$.

In his book MacMahon [26] has provided an elaborate and exten-
sive treatment on partitions of an integer and in an obvious manner has
interpreted $S(m, n; \mu)$ (in fact, a more general case is to be discussed in
the next chapter) in terms of partitions (see [26, Section 10, p. 246]). A
partition of an integer N into at most n parts is a vector $(t_1, \ldots, t_n)$ of r
elements such that

$$0 \leq t_1 \leq t_2 \leq \cdots \leq t_n \qquad \text{and} \qquad t_1 + \cdots + t_n = N.$$

Let $p(N, n)$ denote the set of partitions of N into at most n parts with the
restriction that the ith part $t_i \leq m - \mu n - 1 + (i - 1)\mu, i = 1, \ldots, n$.
As a convention, we may take $p(0, n)$ as the zero vector with n elements.
Letting $d = n(m - \mu n - 1) + \binom{n}{2}\mu$, it is easy to show that

$$\sum_{N=0}^{d} |p(N, n)| = |S(m, n; \mu)|. \tag{1.34}$$

Indeed, MacMahon's book contains a greater wealth of material than
that which is mentioned in Chapter 2.

The main feature of vector representation is that each lattice path is
transformed to an n-dimensional lattice point such that the enumeration involves counting lattice points in a certain polyhedron. Except for
a special case, we have not yet counted paths restricted by a general
boundary or equivalently the set of compositions dominated by an
arbitrary composition. The answer to this question is given in the next
chapter.

7. Diagonal Steps

It is of interest to count paths by introducing diagonal steps and, for
simplicity, to consider the set of paths from $(0, 0)$ to (m, n) having r
diagonal steps (those steps parallel to the line $x = y$) and not touching

the line $x = y$. Denote this set by $L(m, n|r)$. To evaluate $|L(m, n|r)|$ we adopt the "balls into cells" technique—a simple but elegant one—as proposed by Goodman and Narayana [14].

Leaving the diagonal steps apart, the number of paths from $(0, 0)$ to $(m - r, n - r)$ that do not touch $x = y$ is

$$\frac{m - n}{m + n - 2r} \binom{m + n - 2r}{n - r}.$$

Each such path passes through $m + n - 2r$ lattice points, excluding the origin, which for convenience will be referred to as cells. In these cells r diagonal steps, or alternatively balls, can be placed. This method gives rise to all desired paths. The number of ways of placing r similar balls into $m + n - 2r$ distinct cells is

$$\binom{m + n - r - 1}{r}.$$

Hence, we have the following theorem:

THEOREM 6

$$|L(m, n)|r)| = \frac{m - n}{m + n - 2r} \binom{m + n - 2r}{m - r} \binom{m + n - r - 1}{r}$$

$$= \frac{m - n}{m + n - r} \binom{m + n - r}{r, n - r}, \tag{1.35}$$

where

$$\binom{a}{j_1, j_2, \ldots, j_k}$$

is the multinomial coefficient given by

$$\frac{a(a - 1) \cdots (a - \sum_{i=1}^{k} j_i + 1)}{\prod_{i=1}^{k} j_i!}.$$

By the same method the number of unrestricted paths with r diagonal steps is

$$\binom{m + n - r}{r, n - r}. \tag{1.36}$$

Although the "balls into cells" technique can be helpful in some other situations, it has indeed limited applications. For example, when the boundary is the line $x = \mu y$, the technique does not seem to lead us to any solution.

Fray and Roselle considered [13] weighted lattice paths, where each unit of x-step, y-step, and diagonal step is given weight x, y, and z, respectively. In a sense x, y, z act like the probabilities associated with the respective steps. However, the number of weighted paths with m x-steps, n y-steps and r diagonal steps is simply obtained by multiplying $x^m y^n z^r$ to the number of unweighted paths. The use of weighted paths is discussed in an exercise.

Let $f(m, n; r)$ be the number of paths from $(0, 0)$ to (m, n) having r diagonal steps and not touching the line $x = \mu y$, μ being a nonnegative integer. Then

$$f(m, 0; 0) = 1,$$
$$f(m, 0; r) = 0, \qquad r \neq 0,$$
$$f(\mu n, n; r) = 0, \qquad n = 1, 2, \ldots, \quad r = 0, 1, \ldots, n,$$
$$\begin{aligned} f(m, n; r) = {}& f(m - 1, n; r) + f(m, n - 1; r) \\ & + f(m - 1, n - 1; r - 1), \quad m > \mu n, \quad r = 0, 1, \ldots, n, \end{aligned}$$

$$(1.37)$$

which is similar to (1.14).

THEOREM 7

$$f(m, n; r) = \frac{m - \mu n}{m + n - r} \binom{m + n - r}{r, n - r}. \qquad (1.38)$$

The theorem can be proved by a direct verification of (1.37).

Expressions (1.35), (1.36), and (1.38) indicate that these paths could be thought to be in correspondence with certain sets of three-dimensional paths without diagonal steps. A general proposition in this regard involving several types of diagonal steps is given in the next chapter. The reader is suggested to see [12, 30, 37, 38] for paths with diagonal steps.

8. Summary and Concluding Remarks

In this chapter basic methods of counting lattice paths, namely, the reflection principle, penetrating analysis, recurrence, generating function, the inclusion–exclusion principle, and vector representation, are introduced. While some of these are standard techniques in solving counting problems, others reveal the usual combinatorial ingenuity which by its magic wand transforms a seemingly difficult problem into one easily amenable to explicit solution. While vector representation technique will play a significant role in the next chapter, an elaboration of the method of penetrating analysis becomes the major tool in Chapter 3.

Only enumeration of paths restricted by straight line boundaries are dealt with, keeping for the next chapter the situations with general boundaries which involve no new technique but refinements of some of the methods already discussed in this chapter. When the boundary is the line $x = \mu y$ ($\mu \geq 0$, not necessarily an integer), the following solution proposed by Takács [39] is given for completeness:

$$|L(m, n; \mu)| = \binom{m + n - 1}{n} \sum_{j=0}^{n} C_j \binom{n}{j} \bigg/ \binom{m + n - 1}{j}, \qquad (1.39)$$

where C_j $(j = 0, 1, \ldots, n)$ are determined by the recurrence formula

$$\sum_{j=0}^{k} C_j \binom{k}{j} \binom{[k\mu] + k - 1}{j} = 0, \qquad k = 1, 2, \ldots, \qquad (1.40)$$

$[x]$ being the largest integer less than or equal to x. Also, it is equally significant to remark that in [9] more results on path counting were obtained. For example, upper and lower bounds for $|L(m, n; \mu)|$ when $0 \leq \mu < (m/n)$ (μ not necessarily an integer) are given as

$$\frac{m - [\mu n]}{m + n} \binom{m + n}{n} \leq |L(m, n; \mu)| \leq \frac{\min_{(r/s) < \mu} (ms - nr)}{m + n} \binom{m + n}{n}.$$

$$(1.41)$$

Another isolated but curious result is given by Narayana [32] (also see [34]) on lattice paths that do not cross the line $x = \mu y$, μ being a positive integer. Denote by $[n; a_1, \ldots, a_\mu]$ the set of paths $(x_1, \ldots, x_n)$ such that

(a) $x_i \leq \mu_i$, $i = 1, \ldots, n$, and
(b) each vector in $[n; a_1, \ldots, a_\mu]$ has exactly a_j of its positive elements $\equiv j(\mathrm{mod}\ \mu)$, $j = 1, \ldots, \mu$.

Then

$$|[n; a_1, \ldots, a_\mu]| = \sum_{y_1=0}^{a_1} \cdots \sum_{y_\mu=0}^{a^\mu} |[n-1; y_1, \ldots, y_\mu]|$$

$$= \begin{cases} \prod_{i=1}^{\mu} \binom{n+a_i}{a_i} \left(1 - \frac{\sum_{i=1}^{\mu} a_i}{n+1}\right) & \text{if } \sum_{i=1}^{\mu} a_i \le n, \\ 0 & \text{otherwise.} \end{cases}$$

$$(1.42)$$

As an important note one may be reminded of a series of papers by Grossman on lattice paths ([16–23]), which had been considered as a part of recreational mathematics until recently when its impact was seen in various applications. At times Grossman only stated a result without any proof and at others he presented old ideas in a new attractive form. It is no wonder that quite a few of his results were independently rederived later by some other authors.

The classical ballot problem, which notwithstanding its unrealistic formulation as a ballot problem has historcially formed the core for path counting problems and has served as a model for various applied problems, is stated at the beginning and is frequently discussed. The fact that lattice paths have a natural 1:1 correspondence with the compositions of an integer is presented along with the concept of domination which, besides its significance in the context of the present chapter, would play a major role in the general boundary case. The relation of paths with Young tableaux and Fibonacci numbers has been pointed out. At the end paths with simple diagonal steps are enumerated, especially with the aid of the "balls into cells" technique.

Illustrations, rather than a comprehensive treatment, are given to show that a particular problem can be solved by several methods. In the literature generating functions have been used to handle quite a few problems in addition to one mentioned here. Thus no claim is made to an exhaustive account of the subject. The overall development tends to possess an intuitive appeal, which is highlighted in Section 4 on penetrating analysis. Because paths that do not cross certain boundaries can be seen as those which do not touch some other boundaries, discussion of the former is practically omitted. No effort is deliberately made to put forth the relation between lattice paths and various problems in probability and statistics, which discussion is postponed to be treated exclusively in later chapters, in order to emphasize the vital importance of such applications.

The duality concept is introduced by which a result can be written in a different way when either of the two operators, viz., reversion and conjugation (reflection), or both are applied to the paths under consideration. Later, by successfully using them on parts of a path and combining the resulting parts suitably, we shall be able to formulate bijective construction procedures that lead to quite a few interesting and important propositions.

The list of references may not be complete on any particular topic. For instance, the literature contains several papers on ballot problems in various forms, sometimes in a disguised way. Our list generally includes only those references which are directly relevant for the development of a given theme. It also happens that a particular result has been given in several papers and the author has usually chosen only one of these as a reference. In doing so, no claim is made on historical priority. This practice is followed also in other chapters.

One may find some of the exercises quite trivial. The inclusion of such exercises often makes it easier to understand either certain structural aspects of various techniques or steps of simplification.

Exercises

1. Prove (1.4).
2. Derive (1.5) from (1.4).
3. Using (1.4), prove that the number of paths from the origin to $(n, n - 1)$, not touching the line $ny = (n - 1)x$ except at the endpoints, is $(1/(2n - 1))\binom{2n-1}{n}$.
   [Note: It can be obtained from the remark following (1.27).]
4. Show that the number of paths from the origin to $(\alpha_1, \ldots, \alpha_r)$, $\alpha_1 \geq \cdots \geq \alpha_r > 0$, such that the paths do not cross the hyperplanes $x_i = x_{i+1}, i = 1, \ldots, n - 1$, is given by (1.6). If the paths are not to touch the hyperplanes, show that the result is given by [42]:
$$\frac{(\sum_{i=1}^r \alpha_i)!}{\prod_{i=1}^r \alpha_i!} \prod_{i<j} \left(\frac{a_i - a_j}{a_i + a_j}\right).$$
   (Note: An alternative formula is given in [3], by using the reflection principle.)

5. Check (1.8) and hence derive (1.7).

6. When $m = n$ and $t = s$, show that (1.7) is reduced to

$$\sum_{i=-r}^{r} (-1)^i \binom{2n}{n+it},$$

where $r = [n/t]$, the largest integer in n/t.
(See Wilks [45, p. 458].)

7. Consider an urn containing n cards marked with nonnegative integers $k_1, \ldots, k_n$, respectively, such that

$$\sum_{i=1}^{n} k_i = k \leq n.$$

Cards are drawn without replacement from the urn. Let X_r, $r = 1, \ldots, n$, be the random variable representing the number on the card drawn at the rth drawing. Assume that all possible draws are equally likely. Then prove

$$P(X_1 + \cdots + X_r < r \qquad \text{for} \quad r = 1, \ldots, n) = 1 - \frac{k}{n},$$

where $P(\cdot)$ denotes the probability of the event $(\cdot)$, by the method of penetrating analysis [27]. Also show that when $k_1 = \cdots = k_a = 0$, $k_{a+1} = \cdots = k_n = \mu + 1$, the problem reduces to the ballot problem of Theorem 3.

8. By using path interpretation, prove the following convolution identities:

$$\sum_{k=0}^{n} \frac{a}{a+\mu k} \binom{a+\mu k}{k} \frac{b}{b+\mu(n-k)} \binom{b+\mu(n-k)}{n-k}$$
$$= \frac{a+b}{a+b+\mu n} \binom{a+b+\mu n}{n},$$

$$\sum_{k=0}^{n} \binom{a+\mu k}{k} \frac{b}{b+\mu(n-k)} \binom{b+\mu(n-k)}{n-k} = \binom{a+b+\mu n}{n}.$$

9. Check (1.12) and solve for $f_r(k)$ in (1.13).

10. By using the generating function, show that the number of paths from the origin to

$$(x_0, x_1, \ldots, x_r) = \left(\alpha + \sum_{i=1}^{r} \mu_i n_i, n_1, \ldots, n_r\right)$$

which do not touch the hyperplane

$$x_0 = \alpha + \sum_{i=1}^{r} \mu_i x_i$$

except at the end is given by

$$\frac{\alpha}{\alpha + \sum_{i=1}^{r} (\mu_i + 1)n_i} \left(\begin{matrix} \alpha + \sum_{i=1}^{r} (\mu_i + 1)n_i \\ n_1, \ldots, n_r \end{matrix} \right),$$

the μ_is being nonnegative integers and $\alpha > 0$.

(Hint: Use the extended Lagrange inversion formula for power series expansions [15, Section 105].)

11. Letting

$$\phi_k = \sum_{t=1}^{k} \phi_{k,t},$$

prove that

$$\phi_k = \sum \prod_i \frac{F_i^{k_i}}{k_i!},$$

the sum being extended over all k_i such that $\sum i k_i = k$. Write down the recurrence relations for ϕ_k similar to those of the ψ_ks.

12. Prove that (1.31) follows from (1.30).

13. Prove that the solution of the recurrence relation (1.31) is given by

$$D_{k+1} = \left\| \left(\begin{matrix} T_{k-j} + j - 1 \\ j - i + 1 \end{matrix} \right)_+ \right\|_{k \times k},$$

where $\|(a_{ij})\|_{k \times k}$ stands for the kth order determinant whose (i, j)th element is a_{ij}.

14. Prove Lemma 2.

15. Prove (1.34).

16. Prove Theorem 5.

17. Using path combinatorics, prove the identity

$$\sum_{k=0}^{p} \frac{a+k}{a+k+(p-k)(b+1)} \left(\begin{matrix} a+k+(p-k)(b+1) \\ p-k \end{matrix} \right)$$

$$= \frac{a+1}{a+1+p(b+1)} \left(\begin{matrix} a+1+p(b+1) \\ p \end{matrix} \right).$$

18. Prove the identity

$$N_{p,q}(a, b; c-1, 0) + N_{p,q-1}(a, b; c, 0) = N_{p,q}(a, b; c, 0).$$

19. Prove by the "balls into cells" technique that the number of paths from the origin to (m, n, k), $m > n$, with r cube diagonal steps [of the type from (x, y, z) to $(x + 1, y + 1, z + 1)$] such that they do not touch the plane $x = y$ except at the origin is

$$\frac{m - n}{m + n + k - 2r} \binom{m + n + k - 2r}{r, m - r, n - r}.$$

20. Let $g(m, n)$ stand for weighted paths from $(0, 0)$ to (m, n) with diagonal steps and let the generating function of $g(m, n)$ be

$$G(s; n) = \sum_{m=0}^{\infty} g(m, n)s^m.$$

Note that

$$g(m, 0) = x^m \quad \text{and} \quad g(0, n) = y^n.$$

Show that

$$G(s; n) = xsG(s; n) + yG(s; n - 1) + zsG(s, n - 1)$$

and that

$$G(s; n) = (y + zu)^k(1 - xu)^{-(k + 1)}.$$

Hence prove the result in (1.36).

21. Prove that the number of paths from the origin to (m, n, k), $m > \mu n$, with r_1 xy-diagonal steps, r_2 xz-diagonal steps, r_3 yz-diagonal steps, and r_4 cube diagonal steps such that they do not touch the plane $x = \mu y$ (μ a positive integer) except at the origin is [12]

$$\frac{m - \mu n}{m + n + p - w - r_4} \binom{m + n + p - w - r_4}{r_1, r_2, r_3, r_4, m - w + r_3, n - w + r_2},$$

where $w = r_1 + r_2 + r_3 + r_4$. (In [12] similar results exist.)

22. Prove (1.41).
23. Prove (1.42). What happens when $\mu = 1$?

References

1. Alter, R., Some remarks and results on Catalan numbers, *Proc. 2nd Louisiana Conf. on Combinatorics, Graph Theory and Computing, 1971*, pp. 109–132.
2. André, D., Solution directe du problème résolu par M. Bertrand, *C.R. Acad. Sci. Paris* **105** (1887), 436–437.

3. Barton, D. E., and Mallows, C. L., Some aspects of the random sequences, *Ann. Math. Statist.* **36** (1965), 236–260.
4. Bertrand, T., Solution d'un probleme, *C.R. Acad. Sci. Paris* **105** (1887), 369.
5. Bizley, M. T. L., Derivation of a new formula for the number of minimal lattice paths from $(0, 0)$ to (km, kn) having just t contacts with the line; and a proof of Grossman's formula for the number of paths which may touch but do not rise above this line, *J. Inst. Actuar.* **80** (1954), 55–62.
6. Brown, W. G., Historical note on a recurrent combinatorial problem, *Amer. Math. Monthly* **72** (1965), 973–977.
7. Carlitz, L., and Riordan, J., Two element lattice permutation numbers and their q-generalization, *Duke Math. J.* **31**, (1964), 371–388.
8. Carlitz, L., and Roselle, D. P., Triangular arrays subject to MacMahon's conditions, *Fibonacci Quart.* **10** (1972), 591–598.
9. Dvoretzky, A., and Motzkin, T. H., A problem of arrangements, *Duke Math. J* **14** (1947), 305–313.
10. Feller, W., *An Introduction to Probability Theory and Its Applications*, Vol. 1, 2nd ed. Wiley, New York, 1957.
11. Frame, J. S., Robinson, G. De B., and Thrall, R. M., The hook lengths of G_n, *Canad. J. Math.* **6** (1954), 316–325.
12. Fray, R. D., Three dimensional weighted lattice paths, *Proc. 2nd Louisiana Conf. on Combinatorics, Graph Theory and Computing, 1971*, pp. 235–252.
13. Fray, R. D., and Roselle, D. P., Weighted lattice paths, *Pacific J. Math.* **37**, (1971), 85–95.
14. Goodman, E., and Narayana, T. V., Lattice paths with diagonal steps, *Canad. Math. Bull.* **12** (1969), 847–855.
15. Goursat, E., *Functions of a Complex Variable*, Vol. II, Part 1. Dover, New York, 1959.
16. Grossman, H. D., Fun with lattice points—4, 4a, 5, *Scripta Math.* **12** (1946), 223–225.
17. Grossman, H. D., Fun with lattice points—15, *Scripta Math.* **14** (1948), 160–162.
18. Grossman, H. D., Fun with lattice points—16, *Scripta Math.* **15** (1949), 79–81.
19. Grossman, H. D., Fun with lattice points—17, 18, 19, *Scripta Math.* **15** (1949), 232–237.
20. Grossman, H. D., Fun with lattice points—21, *Scripta Math.* **16** (1950), 120–124.
21. Grossman, H. D., Fun with lattice points—22, 23, *Scripta Math.* **16** (1950), 207–212.
22. Grossman, H. D., Fun with lattice points—24, *Scripta Math.* **18** (1952), 298–300.
23. Grossman, H. D., Fun with lattice points—25, *Scripta Math.* **20** (1954), 203–204.
24. Kreweras, G., Sur une classe de problemes de denombrement liés au treillis des partitions des entiers, *Cahiers du Bur. Univ. de Rech. Opér.* **6** (1965), 5–105.
25. Lyness, R. C., Al Capone and the death ray, *Math. Gaz.* **25** (1941), 283–287.
26. MacMahon, P. A., *Combinatory Analysis*, Vols. I and II, Chelsea, Bronx, New York, 1960.
27. Mohanty, S. G., An urn problem related to the ballot problem, *Amer. Math. Monthly* **73** (1966), 526–528.
28. Mohanty, S. G., On a partition of generalized Fibonacci numbers, *Fibonacci Quart.* **6** (1968), 22–33.
29. Mohanty, S. G., and Narayana, T. V., Some properties of compositions and their application to probability and statistics I, *Biometrische Z.* **3** (1961), 252–258.

30. Moser, L., and Zayachkowski, W., Lattice paths with diagonal steps, *Scripta Math.*
    26 (1961), 223–229.
31. Narayana, T. V., A combinatorial problem and its application to probability
    theory I, *J. Indian Soc. Agric. Statist.* **7** (1955), 1969–178.
32. Narayana, T. V., An analogue of the multinomial theorem, *Canad. Math. Bull.* **5**
    (1962), 43–50.
33. Narayana, T. V., and Fulton, G. E., A note on the compositions of an integer,
    Canad. Math. Bull. **1** (1968), 169–173.
34. Narayana, T. V., and Rohatgi, V. K., A refinement of ballot problem, *Skand.*
    Aktuarietidskr. (1965), 222–231.
35. Pólya, G., and Szegö, G., *Aufgaben and Lehrsätze aus der Analysis*, Vol. I. Springer-
    Verlag, Berlin and New York, 1954.
36. Riordan, J., *Combinatorial Identities*, Wiley, New York, 1968.
37. Rohatgi, V. K., A note on lattice paths with diagonal steps, *Canad. Math. Bull.* **7**
    (1964), 470–472.
38. Stocks, D. R., Jr., Lattice paths in E^3 with diagonal steps, *Canad. Math. Bull.* **10**
    (1967), 653–658.
39. Takács, L., A generalization of the ballot problem and its application in the theory
    of queues, *J. Amer. Statist. Assoc.* **57** (1962), 327–337.
40. Takács, L., Ballot Problems, *Z. Wahrsch. Verw. Gebiete* **1** (1962), 154–158.
41. Takács, L., *Combinatorial Methods in the Theory of Stochastic Process.* Wiley,
    New York, 1967.
42. Thrall, R. M. A combinatorial problem, *Michigan Math. J.* **1** (1952), 81–88.
43. Whittaker, E. T., and Watson, G. N., *A Course of Modern Analysis*, 4th ed. Cam-
    bridge Univ. Press, London and New York, 1927.
44. Whitworth, W. A., *Choice and Chance, 4th ed.* Deighton Bell, Cambridge, 1886.
45. Wilks, S. S., *Mathematical Statistics.* Wiley, New York, 1962.
46. Young, A., On quantitative substitutional analysis, *Proc. London Math. Soc.* **28**
    (1927), 255–292.

2 Path Counting—General Boundaries

1. Introduction

The present chapter is a continuation of Chapter 1, where again lattice paths that avoid certain general boundaries are enumerated. From the domination (see Definition 2, Chapter 1) viewpoint the most general two-boundary problem can be described in terms of paths simultaneously dominating $\mathbf{b} = (b_1, \ldots, b_n)$ and dominated by $\mathbf{a} = (a_1, \ldots, a_n)$ with $b_i \leq a_i$ for all i. First, we express the number of paths between two boundaries explicitly by using the notion of domination of vectors, and again obtain another expression for the same number by applying the method of subtraction, which was introduced in conjunction with the reflection principle in the last chapter. Interestingly, each result has the advantage over the other, depending on the boundaries under consideration. Next, we present a general theorem on domination of which the enumeration of paths becomes a special case and study some of its variations and extensions. Also, higher dimensional paths under restrictions are counted. Finally, an important 1 : 1 correspondence is given between paths with various types of diagonal steps (defined later) and paths without diagonal steps. We conclude by providing a continuous analogue of the theorem on domination.

2. Paths within General Boundaries

Let $(\mathbf{b}, \mathbf{a})$ be the set of paths that dominate $\mathbf{b}$ and are dominated by $\mathbf{a}$.

THEOREM 1

$$|(\mathbf{b}, \mathbf{a})| = \det_{n \times n}(c_{ij}) = \|c_{ij}\|_{n \times n}, \qquad (2.1)$$

where

$$c_{ij} = \binom{a_i - b_j + 1}{j - i + 1}_+ \qquad \text{or} \qquad c_{ij} = \binom{a_{n-j+1} - b_{n-i+1} + 1}{j - i + 1}_+ \qquad (2.2)$$

and $\binom{y}{z}_+$ is defined in Theorem 2 of Chapter 1. Note that two forms are obtained by merely arranging the rows and columns of the determinant.

Proof

From the vector viewpoint, $|(\mathbf{b}; \mathbf{a})|$ represents the number of vectors $\mathbf{x} = (x_1, \ldots, x_n)$ such that the x_is are integers satisfying $0 \le x_1 \le \cdots \le x_n$ and $b_i \le x_i \le a_i$ for all i. Therefore

$$|(\mathbf{b}; \mathbf{a})| = \sum_{x_1 = b_1}^{a_1} \sum_{x_2 = y_2}^{a_2} \cdots \sum_{x_n = y_n}^{a_n} 1, \qquad (2.3)$$

where $y_k = \max(b_k, x_{k-1}), k = 2, \ldots, n$. We may replace 1 on the right-hand side by the triangular determinant

$$\left\| \binom{x_{n-j+1} - b_{n-i+1}}{j - i} \right\|.$$

Consider the sum over x_t which appears in the $(n + 1 - t)$th column. If $y_t = x_{t-1}$ (when $t = 1$, this situation does not arise), then

$$\sum_{x_t = x_{t-1}}^{a_t} \binom{x_t - b_{n-i+1}}{n + 1 - t - i}_+ = \binom{a_t - b_{n-i+1} + 1}{n + 2 - t - i}_+$$
$$- \binom{x_{t-1} - b_{n-i+1}}{n + 2 - t - i}_+. \qquad (2.4)$$

The summation over x_t will change only the $(n + 1 - t)$th column of the determinant and make the ith row equal to the right-hand side of (2.4). The negative terms, being equal to the corresponding elements in the next column to the right, drop out of the determinant. On the other hand, if $y_t = b_t$, the negative term is

$$\binom{b_t - b_{n-i+1}}{n + 2 - t - i}_+,$$

which is zero for $i = 1, \ldots, n + 1 - t$ since in that case $b_t \leq b_{n-i+1}$. Also, $b_t > x_{t-1}$ (which follows from $y_t = b_t$) along with $x_{t-1} \geq x_{t-2} \geq \cdots \geq x_1$ imply that each element of the determinant from the first row to the $(n - t + 1)$th row and from the $(n - t + 2)$th column to the last column is zero. It follows from the properties of the determinant that the negative terms in this case also drop out. When we take the summation over the consecutive columns starting from the first, we conclude (2.1). This completes the proof.

The result which is a special case of a general one to be given later (see Kreweras [8]) is obtained in this form by Steck [24], in connection with the distribution of two-sample Kolmogorov–Smirnov statistics. Though it would not be difficult to get a proof by the method of inclusion–exclusion (see [21], [23]), the present method of exploitation of simple properties of determinants is given in [13], as an adaptation of Narayana's proof [18]. It has the advantage that Kreweras' general case can be proved with the same amount of ease as we have witnessed above. See also Kryscio [9], whose interest in the problem stems from an altogether different situation.

Theorem 1 provides an answer to the composition problem that the number of r-compositions of N which dominate $(t'_1, \ldots, t'_r)$ and are dominated by $(t_1, \ldots, t_r)$ is

$$\left\| \binom{\sum_{k=1}^{r-j} t_k - \sum_{k=1}^{r-i} t'_k + j - i + 1}{j - i + 1}_+ \right\|_{(r-1) \times (r-1)}. \tag{2.5}$$

To find the number D_{r-1} (see the notation of Ds and Ts in Lemma 1, Chapter 1) of r-compositions of N which are dominated by $(t_1, \ldots, t_r)$, we have to put $t_1' = \cdots = t_{r-1}' = 1$ and get

D_{r-1}

$$
= \begin{vmatrix}
\binom{T_{r-1}-r+2}{1}_+ & \binom{T_{r-2}-r+3}{2}_+ & \cdots & \binom{T_2-1}{r-2}_+ & \binom{T_1}{r-1}_+ \\
1 & \binom{T_{r-2}-r+3}{1}_+ & \cdots & \binom{T_2-1}{r-3}_+ & \binom{T_i}{r-2}_+ \\
0 & 1 & \cdots & \binom{T_2-1}{r-4}_+ & \binom{T_1}{r-3}_+ \\
\vdots & \vdots & \vdots & \vdots & \vdots \\
0 & 0 & \cdots & 1 & \binom{T_1}{1}_+
\end{vmatrix}.
$$

$$(2.6)$$

For an alternative form of D_r see Exercise 3. Expanding D_{r-1} by its first column, one gets

$$
D_{r-1} = \binom{T_{r-1}-r+2}{1}_+ D_{r-2} - R_{r-2},
$$

where R_{r-2} is D_{r-1} with column 1 and row 2 deleted. In a similar manner we have

$$
R_{r-2} = \binom{T_{r-2}-r+3}{2}_+ D_{r-3} - R_{r-3}.
$$

Repeated expansion of $R_{r-2}, R_{r-3}, \ldots$ in terms of its first column yields (1.31). The entire result could be written in the language of vectors, instead of compositions.

Carlitz et al. [3] observed that the one-boundary case has an interesting property which apparently arose in a class of convolution codes in a paper by Berlekamp [2]. Here we are considering paths dominated by the path $\mathbf{a} = (a_1, \ldots, a_n)$. Path $\mathbf{a}$ determines a_i cells in the ith row $i = 1, \ldots, n$. Let the (i, j)th cell be the one in the ith row from the top and jth column from the right. Let c_{ij} be assigned to the (i, j)th cell, $j = 1, \ldots, a_i, i = 1, \ldots, n$, such that $c_{1j} = 1$ $(j = 1, \ldots, a_1)$, $c_{i, a_i} = 1$ $(i = 2, \ldots, n)$, and c_{ij} in general is the number of paths dominated by $((a_1 - j)_+, \ldots, (a_{i-1} - j)_+)$ with $(x)_+ = \max(0, x)$. Then the property of these arrays of c_{ij}s is that every square subarray comprising adjacent rows and columns and containing at least one 1 will have determinant 1 (see Exercise 5). We call this property B (after Berlekamp).

5	4	3	2	1	j $\diagdown$ i
				1	1
			1	1	2
1	1	1	2	5	3
1	2	3	7	19	4

This was generalized by Fray and Roselle [4] for weighted paths with diagonal steps.

As an illustration let $\mathbf{a} = (2, 3, 5, 5)$. The accompanying table gives c_{ij} entries. The entry c_{41} is the number of paths dominated by $(1, 2, 4)$, which is equal to 19. Other entries are similarly obtained. The stated property can easily be checked to be true.

We have noticed in the proof of Theorem 1 that the path counting problem is geometrically equivalent to counting n-dimensional lattice points in a polyhedron determined by hyperplanes $x_i = b_i, x_i = a_i, i = 1, \ldots, n$, and $x_i = x_{i+1}, i = 1, \ldots, n - 1$. Instead of transforming paths into lattice points, we may continue our original technique (Chapter 1, Section 3) of subtracting the number of paths that are undesirable from the total number of paths. However, the main obstacle is to find an explicit expression for the number of undesirable paths, which was calculated earlier by the use of the reflection principle.

Let us first consider the one-boundary case with boundary path $\mathbf{a}$. For simplicity in writing let $|(\mathbf{a})| = |(\mathbf{0}, \mathbf{a})|$. It is easily checked that

$$|(\mathbf{a})| = \binom{a_n + n}{n} - N^*(a_1, \ldots, a_n), \qquad (2.7)$$

where $N^*(a_1, \ldots, a_n)$ denotes the number of paths that cross the boundary.

Also

$$N^*(a_1) = 0,$$

$$N^*(a_1, a_2) = \binom{a_2 - a_1 + 1}{2}_+,$$

$$N^*(a_1, \ldots, a_n) = \binom{a_n - a_1 + n - 1}{n}_+ \qquad (2.8)$$

$$+ \sum_{i=2}^{n-1} \binom{a_n - a_i + n - i}{n - i + 1}_+ |(a_1, \ldots, a_{i-1})|$$

$$\text{for} \quad n > 2,$$

yielding

$$|(a_1)| = \binom{a_1 + 1}{1},$$

$$|(a_1, a_2)| = \binom{a_2 + 2}{2} - \binom{a_2 - a_1 + 1}{2}_+,$$

$$|(a_1, \ldots, a_n)| = \binom{a_n + n}{n} - \binom{a_n - a_1 + n - 1}{n}_+ \qquad (2.9)$$

$$- \sum_{i=2}^{n-1} \binom{a_n - a_i + n - i}{n - i + 1}_+ |(a_1, \ldots, a_{i-1})|$$

$$\text{for} \quad n > 2,$$

which is the same as (9) in [5] and (2) in [26]. (In [5], Göbel's relation (3) is also equivalent to Lemma 1 of Chapter 1.) The explicit solution of (2.9) is stated as a determinant.

THEOREM 2

$$|(\mathbf{a})| = \det_{n \times n} (d_{ij}) = \|d_{ij}\|_{n \times n}, \qquad (2.10)$$

where

$$d_{ij} = \begin{cases} \binom{a_n + n}{n} & \text{when} \quad i = j = 1, \\[2ex] \binom{a_{n-i} + n - i}{n - i}_+ & \text{when} \quad j = 1, \quad i \neq 1, \\[2ex] \binom{a_n - a_{n-j+1} + j - 1}{j}_+ & \text{when} \quad i = 1, \quad j \neq 1, \\[2ex] \binom{a_{n-1} - a_{n-j+1} + j - i - 1}{j - i}_+ & \text{when} \quad i \neq 1, \quad j \neq 1. \end{cases}$$

Proof

For $n = 1, 2$ the determinant checks with (2.9). We have only to verify that it satisfies the recurrence relation in (2.9). Notice that the d_{1j}s are the binomial coefficients on the right-hand side of (2.9).

Form a new determinant $D_{(n+1) \times (n+1)}$ from (2.10) by

(a) putting a column between the first and second, with all elements equal to zero except the second one which is equal to one, and
(b) putting a row between the first and the second with elements

$$\binom{a_{n-1} + n - 1}{n - 1}_{+}, \ 1, \ \binom{a_{n-1} - a_{n-1}}{1}_{+},$$

$$\binom{a_{n-1} - a_{n-2} + 1}{2}_{+}, \ \ldots, \ \binom{a_{n-1} - a_1 + n - 2}{n - 1}_{+}.$$

Clearly, $D_{(n+1) \times (n+1)} = \|d_{ij}\|_{n \times n}$ for $n \geq 2$.

The expansion of $\|d_{ij}\|_{n \times n}$ by the first row shows that the cofactor of

$$\binom{a_n - a_k + n - k}{n - k + 1}_{+}$$

after some simplification is

$$D_{k \times k} = \|d_{ij}\|_{(k-1) \times (k-1)} = |(a_1, \ldots, a_{k-1})|.$$

Therefore the recurrence relation is satisfied, and this completes the proof.

When $a_n = a_{n-1} = \cdots = a_{n-r} > a_{n-r-1}$, the determinant is reduced to one of order $n - r$. In the extreme case we have the well-known result

$$|(a, a, \ldots, a)| = \binom{a + n}{n} \tag{2.11}$$

since each element above the main diagonal of the determinant is zero. One striking feature in the general boundary situation is the determinantal form of the explicit solution. When (2.10) is compared with the similar formula obtained from Theorem 1, one gets a determinantal identity. Convenience determines the choice between the two forms for computational purposes. Evidently, if the boundary is away from the center, formula (2.10) is preferred. In fact the product of elements in the main diagonal by Theorem 1 is

$$\prod_{i=1}^{n} \binom{a_i + 1}{1}$$

whereas by Theorem 2 it is

$$\binom{a_n + n}{n}.$$

The one-boundary case has been dealt with by the author in [11, 12]. Some simplifications of the determinant are considered in [24]. A not too obvious identity arises from the remark following (1.27) and is given below.

$$\left|\left(\left[\frac{m}{n}\right], \left[\frac{2m}{n}\right], \ldots, \left[\frac{(n-1)m}{n}\right]\right)\right| = \frac{1}{m+n}\binom{m+n}{n}.$$

In case of two boundaries, say A and B, we adopt the same procedure; namely, from the total number of paths which do not cross B subtract the number of paths which do not cross B but cross A. Observe that the one-boundary case is equivalent to the situation with two boundaries when boundary B has vector representation $(0, 0, \ldots, 0)$. The next result for two boundaries is stated without proof:

$$|(\mathbf{b}, \mathbf{a})| = \|d_{ij}^*\|_{n \times n}, \tag{2.12}$$

where

$$d_{ij}^* = \begin{cases} |(a_n - b_n, a_n - b_{n-1}, \ldots, a_n - b_1)| & \text{for } i = j = 1, \\ |(a_{n-i} - b_{n-i}, a_{n-i} - b_{n-i}, \ldots, \\ \quad a_{n-i} - b_1)| & \text{for } j = 1, \ i \neq 1, n, \\ |(h(a_n - b_n, a_n - a_{n-j+1} - 1), \\ \quad h(a_n - b_{n-1}, a_n - a_{n-j+1} - 1), \\ \quad \ldots, h(a_n - b_{n-j+1}, \\ \quad a_n - a_{n-j+1} - 1))| & \text{for } i = 1, \ j \neq 1, \\ |(h(a_{n-i} - b_{n-i}, a_{n-i} - a_{n-j+1} - 1), \\ \quad h(a_{n-i} - b_{n-i-1}, \\ \quad a_{n-i} - a_{n-j+1} - 1), \\ \quad \ldots, h(a_{n-i} - b_{n-j+1}, \\ \quad a_{n-i} - a_{n-j+1} - 1))| & \text{for } i \neq 1, \ i < j, \\ 1 & \text{for } i = j, \ i \neq 1, \\ 1 & \text{for } i = n, \ j = 1, \\ 0 & \text{for } i > j, \ j \neq 1, \end{cases}$$

and

$$h(\cdot, \cdot) = \min(\cdot, \cdot).$$

Some simplifications of the determinant in (2.12) are possible under special situations, the trivial one being the reduction to the one-boundary case.

Also, if $b_n \leq a_1$, we get

$$d_{ij}^* = \binom{a_n - a_{n-j+1} + j - 1}{j}_+, \qquad j \neq 1,$$

and

$$d_{ij}^* = \binom{a_{n-i} - a_{n-j+1} + j - i - 1}{j - i}_+, \qquad i < j, \quad i \neq 1.$$

Another interesting case arises when no path can cross both boundaries. In such an event a simple argument yields

$$|(\mathbf{b}, \mathbf{a})| = |(\mathbf{a})| + |(a_n - b_n, a_n - b_{n-1}, \ldots, a_n - b_1)| - \binom{a_n + n}{n}.$$

$$(2.13)$$

The consideration of duality (say, by interchanging A and B) gives rise to alternative expressions for (2.12). The main feature of the present approach is to write recursive relations that are exhibited in (2.8) and (2.9). A routine application of the method would enable us to handle the counting of higher dimensional paths under certain specific restrictions.

3. Some Generalizations

A combinatorial theorem based on domination (Kreweras [8, p. 64]), which generalizes Theorem 1, is now given.

Let $s^{(j)} = (s_{1j}, \ldots, s_{nj}), j = 1, 2, \ldots$, be a set of vectors with integer-valued elements satisfying

$$0 \leq s_{1j} \leq \cdots \leq s_{nj},$$
$$b_i \leq s_{ij} \leq s_{i, j+1} \leq a_i \qquad (i = 1, \ldots, n; \quad j = 1, 2, \ldots). \quad (2.14)$$

Denote by $|(\mathbf{b}, \mathbf{a}; r)|$ the number of $n \times r$ matrices $[s_{ij}]$ satisfying (2.14).

THEOREM 3

For $r = 1, 2, \ldots$

$$|(\mathbf{b}, \mathbf{a}; r)| = \|c_{ij}^{(r)}\|_{n \times n}, \qquad (2.15)$$

where

$$c_{ij}^{(r)} = \binom{a_i - b_j + r}{r + j - i}_+ \quad \text{or} \quad c_{ij}^{(r)} = \binom{a_{n-j+1} - b_{n-i+1} + r}{r + j - i}_+.$$

(2.16)

Proof

Clearly $|(\mathbf{b}, \mathbf{a}; 1)| = |(\mathbf{b}, \mathbf{a})|$, and the result checks. Assume that the theorem is true for all integers up to $(r - 1)$, so that

$$|(\mathbf{b}, \mathbf{a}; r - 1)| = \left\| \binom{a_{n-j+1} - b_{n-i+1} + r - 1}{r + j - i - 1}_+ \right\|.$$

(2.17)

It is easily seen that

$$|(\mathbf{b}, \mathbf{a}; r)| = \sum_{s_{1r} = b_1}^{a_1} \sum_{s_{2r} = u_2}^{a_2} \cdots \sum_{s_{nr} = u_n}^{a_n} |(\mathbf{b}, \mathbf{s}^{(r)}; r - 1)|$$

(2.18)

with $u_k = \max(b_k, s_{k-1,r})$, $k = 2, \ldots, n$. Note that (2.18) is similar to (2.3). When the summation is performed successively by using the identity

$$\sum_{s_{tr} = u_t}^{a_t} \binom{s_{tr} - b_{n-i+1} + r - 1}{r + n - t - i}_+$$
$$= \binom{a_t - b_{n-i+1} + r}{r + n - t - i + 1} - \binom{u_t - b_{n-i+1} + r - 1}{r + n - t - i + 1}_+,$$

(2.19)

we observe that the contribution by the negative terms in (2.19) drops out either if $s_{t-1,r} \geq b_t$ or if $b_t > s_{t-1,r}$ $(t \geq 2)$. The same is true for $t = 1$. Thus the conclusion of the theorem follows and the proof is complete.

Besides a great body of significant materials, Kreweras has formulated the theorem in terms of Young's lattice (also see [1, p. 69]), which has a natural geometric interpretation. He has connected it with the standard Young tableaux of Chapter 1. Define a lattice surface as a surface consisting of points which have at least one coordinate as an integer. Thus a two-dimensional lattice surface is only a lattice path. We have considered paths which are representable by nondecreasing nonnegative integer-valued vectors $(x_1, \ldots, x_n)$, where x_i is the distance of the path from $(m, n - i)$, $m \geq x_n$. In other words,

$$m - x_{n-i+1} = \min\{u_i : (u_i, i) \text{ is a point on the path}\}, \qquad i = 1, \ldots, n.$$

Here $x = m$ may be considered as the reference line. Analogously, we consider lattice surfaces which are represented by the $n \times r$ matrices $[s_{ij}]$, satisfying the property that the s_{ij}s are nonnegative integers and

$$s_{ij} \leq s_{i,j+1} \qquad (j = 1, \ldots, r-1, i = 1, \ldots, n),$$
$$s_{ij} \leq s_{i+1,j} \qquad (i = 1, \ldots, n-1; j = 1, \ldots, r),$$

where

$$m - s_{n-i+1,r-j+1} = \min\{u_{ij}: (u_{ij}, i, j) \text{ is a point on the surface}\}$$
$$i = 1, \ldots, n, j = 1, \ldots, r, m \geq s_{nr}$$

(i.e., $x = m$ is the reference plane). Then $|(\mathbf{b}, \mathbf{a}; r)|$ gives the number of such lattice surfaces which do not cross surfaces represented by matrices $[b_{ij}]$ and $[a_{ij}]$ (or in the language of domination, which dominate $[b_{ij}]$ and are dominated by $[a_{ij}]$), with $b_{ij} = b_i$ and $a_{ij} = a_i$ for all i and j. An alternative interpretation similar to that of Theorem 1 would be that if a particle moves in n-dimensional space from the lattice point $\mathbf{b}$ to $\mathbf{a}$ through lattice points stopping at r intermediate points $s^{(1)}, \ldots, s^{(r)}$ satisfying (2.14), we can count in how many different ways this is possible.

Another remark is that in MacMahon's book [10, Vol. II] one finds the expression for the generating function of $|(\mathbf{0}, \mathbf{a}; r)|$ in addition to other interesting results. There exists a great deal of wealth in [10], but in a somewhat obscure fashion.

A rather interesting and useful variation of Theorem 3 is the following. A particle starting from the point $(b_1, \ldots, b_n)$, $0 \leq b_1 \leq \cdots \leq b_n$, reaches the point $(v_1, \ldots, v_n)$ in k stages, where $1 \leq k \leq v_1 \leq \cdots \leq v_n$, according to the following scheme. Let $a_{i\alpha} \geq 1$ be the increase in the ith coordinate at stage α. Then the a_αs satisfy

$$j \leq \sum_{u=1}^{j} a_{1u} \leq \sum_{u=1}^{j} a_{2u} \leq \cdots \leq \sum_{u=1}^{j} a_{nu}, \qquad j = 1, \ldots, k. \quad (2.20)$$

To enumerate all paths let us put

$$\sum_{u=1}^{j} a_{iu} - j = s_{ij}, \qquad i = 1, \ldots, n, \quad j = 1, \ldots, k-1.$$

We immediately notice that we are in the situation of Theorem 3 with $a_i = v_i - k, i = 1, \ldots, n$ and $r = k - 1$, and thus the required number is

$$\left\| \binom{v_{n-j+1} - b_{n-i+1} - 1}{k - 1 + j - i} \right\|_{+ \, n \times n} = \left\| \binom{v_i - b_j - 1}{k - 1 + j - i} \right\|_{+ \, n \times n}. \quad (2.21)$$

Without the bs the above problem was formulated by Narayana [19].

When $n = 2$ and $v_1 = v_2 = n$, the expression in (2.21) is reduced to $(1/n)\binom{n}{k}\binom{n}{k-1}$ (see Narayana [17]), which is also encountered in Riordan's book [21, p. 17]. It represents the number of paths from the origin to (n, n) not crossing the line $x = y$ such that each path has exactly k horizontal and k vertical components (a component is a sequence of steps of the same type followed and preceded by steps of the other type). Using a modified definition of domination [16] when $n = 2$, the preceding variation allows us to rederive the ballot theorem of Chapter 1. (See Exercise 11.)

Apart from this variation one can generate a class of determinantal identities from Theorem 3, by the duality principle introduced in Chapter 1, Section 4. While the consideration of reverse paths gives the trivial identity suggested by alternative expressions in (2.16), we shall quote the other type of identities arising out of conjugation. For simplicity of consideration take $a_1 = \cdots = a_n = m$ and $b_1 = \cdots b_n = 0$, so that the number of paths from $(0, 0)$ to (m, n) as determined by the theorem with $r = 1$ is

$$\left\|\binom{m + 1}{1 + j - i}_+\right\|_{n \times n}.$$

The set of paths from $(0, 0)$ to (n, m) are conjugate to those leading to (m, n). Because of the bijectivity due to conjugation, we have the identity

$$\left\|\binom{m + 1}{1 + j - i}_+\right\|_{n \times n} = \left\|\binom{n + 1}{1 + j - i}_+\right\|_{m \times m} = \binom{m + n}{n}. \qquad (2.22)$$

Once the motivation is clear, we write the construction of conjugation more precisely, so that the identity for general r readily follows. For notational simplicity put $a_n = m$. To each vector $s^{(j)}$ in Theorem 3 associate $u^{(j)} = (u_{1j}, \ldots, u_{mj})$, where

$$u_{ij} = \text{the number of } s_{kj} \text{ less than } i, i = 1, \ldots, m. \qquad (2.23)$$

Clearly the mapping is bijective, and therefore we may say that $s^{(j)}$ and $u^{(j)}$ form a pair of conjugate vectors. Furthermore,

$$0 \le u_{1j} \le u_{2j} \le \cdots \le u_{mj},$$

and if by construction (2.23)

$$\mathbf{a} \Leftrightarrow \mathbf{c} = (c_1, \ldots, c_m) \qquad \text{and} \qquad \mathbf{b} \Leftrightarrow \mathbf{d} = (d_1, \ldots, d_m),$$

then

$$\{b_i \le s_{ij} \le s_{i,j+1} \le a_i, i = 1, \ldots, n, j = 1, 2, \ldots\}$$
$$\Leftrightarrow \{d_i \ge u_{ij} \ge u_{i,j+1} \ge c_i, i = 1, \ldots, m, j = 1, 2, \ldots\}.$$

Thus

$$|(\mathbf{b}, \mathbf{a}; r)| = |(\mathbf{c}, \mathbf{d}; r)|,$$

leading to the class of identities

$$\left\| \binom{a_i - b_j + r}{r + j - i}_+ \right\|_{n \times n} = \left\| \binom{d_i - c_j + r}{r + j - i}_+ \right\|_{m \times m}. \tag{2.24}$$

The question of whether or not Theorem 3 can be generalized further may be asked. We give the formulation of two generalizations. Finally, consider $k \times r$ matrices $[s_{ij}]$ such that

(i) $b_i \leq s_{i1} \leq \cdots \leq s_{ir} \leq a_i, 0 \leq b_i \leq a_i, i = 1, \ldots, k,$

(ii) $s_{ij} \leq s_{i+1,j} + l_i, a_i \leq a_{i+1} + l_i, b_i \leq b_{i+1} + l_i,$
$i = 1, \ldots, k - 1, j = 1, \ldots, r.$

The number of such matrices is

$$\| C_{ij}^{(r)} \|_{k \times k}, \tag{2.25}$$

where

$$C_{ij}^{(r)} = \begin{cases} \dbinom{a_{k-j+1} - b_{k-i+1} - l_{k-i} - \cdots - l_{k-j+1} + r}{r + j - i}_+ & \text{when} \quad i < j, \\[3em] \dbinom{a_{k-j+1} - b_{k-i+1} + l_{k-j} + \cdots + l_{k-i+1} + r}{r + j - i}_+ & \text{when} \quad i > j, \\[3em] \dbinom{a_{k-j+1} - b_{k-j+1} + r}{r}_+ & \text{when} \quad i = j. \end{cases}$$

With $b_i = 0$ and $l_i \geq 0$ for all i the problem was given a different formulation in terms of compositions by Narayana [18]. However he has not given a complete proof of the result. Another generalization is to determine the number of $r \times n$ matrices $[s_{ij}]$ such that

(i) $0 \leq s_{i1} \leq \cdots \leq s_{in}, \quad i = 1, \ldots, r,$

(ii) $b_j \leq s_{1j}, s_{rj} \leq a_j, \quad j = 1, \ldots, n,$

$$0 \leq b_1 \leq \cdots \leq b_r, \qquad 0 \leq a_1 \leq \cdots \leq a_r,$$

(iii) $s_{ij} \leq s_{i+1,j} + d_i, \quad i = 1, \ldots, r - 1, \quad j = 1, \ldots, n.$

If $d_1 \leq b_1, d_1 + d_2 \leq b_1, \ldots, d_1 + \cdots + d_{r-1} \leq b_1$, the number is

$$\left\| \binom{a_{n-j+1} - b_{n-i+1} + \sum_{k=1}^{r-1} d_k + r}{r + j - i}_+ \right\|. \tag{2.26}$$

The second formulation is due to the author of [14], but unfortunately the conditions are wrongly stated. Note that if $l_1 = \cdots = l_{n-1} = 0$ and $d_1 = \cdots = d_r = 0$, the problems are equivalent to the problem in Theorem 3. In either case the technique of the proof is the same as in Theorem 3 in which the elementary properties of a determinant have been employed. The last problem has an unconditional solution in terms of a large determinant obtained by Chorneyko and Zing† under a simpler restriction on the ds, viz., the ds are nonnegative. This is given in Exercise 13(b). However the emphasis in (2.26) is to retain the earlier determinantal form which has necessitated the restrictions on the ds.

Note that these problems can be seen as multi-dimensional path counting problems.

4. Higher Dimensional Paths‡

Until now we have been depending on some elementary properties of the determinant for the solution of our enumeration problems. Recall that we have established the recurrence relation (1.30) [or (1.31)], from which the solution of the enumeration can be worked out (see Exercise 13, Chapter 1). One way of achieving this is by getting an orthogonal relation between two triangular matrices from the recurrence relation, which in turn leads to the explicit solution (to be fully developed below). For two-dimensional paths, this technique is suggested in [3]. For a detailed discussion on orthogonal relations we may refer to Chapter 5. It seems that this approach has an equally natural appeal which enables us to count higher dimensional paths between two boundaries. For this purpose we represent a higher dimensional path by a string of vectors, analogous to the vector representation of a two-dimensional path (Chapter 1, Section 6).

In $(k + 1)$-dimensional space with axes $X_0, X_1, \ldots, X_k$ consider lattice paths from the origin to $(n_0, n_1, \ldots, n_k)$, $n_i \geq 0$ for all i. By the $\mathbf{r} (=(r_1, \ldots, r_k))$th level, where $r_i = 0, 1, \ldots, n_i$, $n_i \geq 0$, $i = 1, \ldots, k$, we mean the set of points $\{(x_0, n_1 - r_1, \ldots, n_k - r_k): 0 \leq x_0 \leq n_0\}$. We say that a path reaches the $\mathbf{r}$th level if it passes through one of the points in that level. Note that a path may or may not reach a particular level.

---

† L. Zing informed me of the result, which was part of her Ph.D. thesis, submitted to McMaster University in 1978.

‡ The result of this section was obtained recently by the author jointly with B. R. Handa. See *Discrete Math.* **26** (1979), 119–128.

Also if it reaches the rth level, it may pass through several of its points before reaching either one of the levels $(r_1, \ldots, r_{i-1}, r_i - 1, r_{i+1}, \ldots, r_k)$, $i = 1, \ldots, k$. Let

$$x(\mathbf{r}) = \begin{cases} n_0 - \min\{x_0\} & \text{if the path reaches the } r\text{th} \\ & \text{level and } (x_0, n_1 - r_1, \ldots, n_k - r_k) \\ & \text{is a point on the path,} \\ 0 & \text{otherwise.} \end{cases}$$

Clearly $0 \le x(\mathbf{r}) \le n_0$ and $x(\mathbf{n}) = n_0$.

The representation of a path is done inductively. We know that a two-dimensional path from the origin to (n_0, n_1) has a nondecreasing vector representation which will be slightly modified due to the new notation. The modified vector notation representation is $(x(0), x(1), \ldots, x(n_1))$, such that referring to Chapter 1, Section 6, we have $n_1 = n$, $x(0) = x_1, x(1) = x_2, \ldots, x(n_1 - 1) = x_n$, $x(n_1) = n_0$. Now consider three dimensional paths. Since such a path from the origin to (n_0, n_1, n_2) intersects with the plane $X_1 = i, i = 0, 1, \ldots, n_1$, in a two-dimensional path (including a degenerate one, viz., a point), it consists of a string of $(n_1 + 1)$ two-dimensional path segments successively lying in planes $X_1 = i, i = 0, 1, \ldots, n_1$, such that the initial point of the path segment on each of the planes is joined by the terminal point of the path segment in the preceding plane by a step along the X_1-axis. Thus any path can be uniquely represented by the matrix

$$\begin{pmatrix} x(0,0) & \cdots & x(0, j_1) & & & \\ & x(1, j_1) & \cdots & x(1, j_2) & & 0 \\ & & x(2, j_2) & \cdots & & \\ & & & & \ddots & \\ 0 & & & & x(n_1, j_{n_1}) & \cdots & x(n_1, n_2) \end{pmatrix}$$

with $x(n_1, n_2) = n_0$, $0 = j_0 \le j_1 \le \cdots \le j_{n_1} \le j_{n_1+1} = n_2$, where the vector $(x(i, j_i), x(i, j_i + 1), \ldots, x(i, j_{i+1}))$ in the $(i + 1)$th row corresponds to the two-dimensional path segment on the plane $X_1 = n_i - i$. Alternatively, we observe that to a path there corresponds a nondecreasing vector with $(n_1 + n_2 + 1)$ nonnegative integer elements arranged along a path from $(0, 0)$ to (n_1, n_2). By an extension of the preceding representation, it is not difficult to see that to each path from the origin to $(n_0, n_1, \ldots, n_k)$ there corresponds a unique nondecreasing vector with $(n_1 + \cdots + n_k + 1)$ nonnegative integer elements arranged along a unique path from the origin to $(n_1, \ldots, n_k)$. If the path in the

$(k + 1)$-dimension passes through the $\mathbf{r}$th level, then the $(r_1 + \cdots + r_k + 1)$th element is $x(\mathbf{r})$.

To consider restricted paths, we denote by $a(\mathbf{r})$ and $b(\mathbf{r})$ the upper and lower restrictions, respectively, at the $\mathbf{r}$th level such that the path at the $\mathbf{r}$th level can pass through only points in the set

$$\{(x_0, n_1 - r_1, \ldots, n_k - r_k); 0 \le b(\mathbf{r}) \le n_0 - x_0 \le a(\mathbf{r}) \le n_0\}.$$

The sets

$$A(\mathbf{n}) = \{a(\mathbf{r}): 0 \le r_i \le n_i, i = 1, \ldots, k\}$$

and

$$B(\mathbf{n}) = \{b(\mathbf{r}); 0 \le r_i \le n_i, i = 1, \ldots, k\}$$

are called, respectively, the upper and lower restrictions on the path. Obviously, we have to set $a(\mathbf{n}) = n_0$ and $b(\mathbf{0}) = 0$, where $\mathbf{0}$ is the k-vector with all zeros. In the following we assume that the restrictions on a path are such that for every $\mathbf{r}$, $a(\mathbf{r})$ and $b(\mathbf{r})$ are nondecreasing in any coordinate; i.e.,

$$a(r_1, \ldots, r_k) \le a(r_1, \ldots, r_{j-1}, r_j + 1, r_{j+1}, \ldots, r_k)$$

and

$$b(r_1, \ldots, r_k) \le b(r_1, \ldots, r_{j-1}, r_j + 1, r_{j+1}, \ldots, r_k)$$

for $j = 1, \ldots, k$. In the two-dimensional case these restrictions are trivially satisfied for any two general boundaries. However in higher dimensional situations there exist boundaries which need not satisfy the above nondecreasing property.

For simplicity and clarity we only consider paths with an upper restriction since we can see that the case with both restrictions does not entail any special difficulty. In this case observe that $0 \le x(\mathbf{r}) \le a(\mathbf{r})$. Denote by $g_k(A(\mathbf{n}))$ the number of paths from the origin to $\mathbf{n}$ with the upper restriction or simply the restrictions $A(\mathbf{n})$. In the next theorem we give a recurrence relation on $g_k(A(\mathbf{n}))$ which is vital for the enumeration of paths.

THEOREM 4

$$\sum_{0 \le \mathbf{r} \le \mathbf{n}} (-1)^{(\mathbf{n}-\mathbf{r}) \cdot \mathbf{1}} \binom{a(\mathbf{r}) + 1}{\mathbf{n} - \mathbf{r}} g_k(A(\mathbf{r})) = \delta_{\mathbf{0}}^{\mathbf{n}}, \qquad (2.27)$$

where $\mathbf{1}$ is the k-vector with all ones,

$$\binom{a(\mathbf{r}) + 1}{\mathbf{n} - \mathbf{r}} = \binom{a(\mathbf{r}) + 1}{n_1 - r_1, \ldots, n_k - r_k},$$

$$\delta_{\mathbf{m}}^{\mathbf{n}} = \begin{cases} 1 & \text{when } \mathbf{m} = \mathbf{n}, \\ 0 & \text{otherwise} \end{cases}$$

(i.e., $\delta_{\mathbf{m}}^{\mathbf{n}}$ is the Kronecker delta), and the vector ordering $\mathbf{x} \leq \mathbf{y}$ means $x_i \leq y_i$ for every i.

Proof

We give the proof for $k = 3$, since the general proof is analogous except that it involves complicated notations. Recall that $g_3(A(n_1, n_2, n_3))$ represents the number of nondecreasing $(n_1 + n_2 + n_3 + 1)$-vectors arranged along three-dimensional paths from the origin to (n_1, n_2, n_3), subject to the restriction that

$$0 \leq x(r_1, r_2, r_3) \leq a(r_1, r_2, r_3), \qquad r_i = 0, 1, \ldots, n_i, \quad i = 1, 2, 3.$$

For a given arrangement in a three-dimensional path the number of nondecreasing $(n_1 + n_2 + n_3 + 1)$-vectors is $g_1(A(n_1 + n_2 + n_3 + 1))$, where

$$A(n_1 + n_2 + n_3 + 1) = \{a(s, t, u): s = 0, 1, \ldots, n_1,$$
$$t = i_s, i_s + 1, \ldots, i_{s+1},$$
$$u = j_{s+t}, j_{s+t} + 1, \ldots, j_{s+t+1},$$
$$i_0 = j_0 = 0, i_{n_1 + 1} = n_2, j_{n_1 + n_2 + 1} = n_3\}.$$

Note that by fixing $(i_1, \ldots, i_{n_1})$ and $(j_1, \ldots, j_{n_1 + n_2})$ we get the arrangement on a particular three-dimensional path. Thus when we vary both sets $(i_1, \cdots, i_{n_1})$ and $(j_1, \ldots, j_{n_1 + n_2})$ such that $0 \leq i_1 \leq \cdots \leq i_{n_1} \leq n_2$ and $0 \leq j_1 \leq \cdots \leq j_{n_1 + n_2} \leq n_3$, we get all arrangements of $(n_1 + n_2 + n_3 + 1)$-vectors along three-dimensional paths. The above discussion leads to

$$g_3(A(n_1, n_2, n_3)) = \sum_{R_1} \sum_{R_2} g_1(A(n_1 + n_2 + n_3 + 1)), \qquad (2.28)$$

where

$$R_1 = \{(i_1, \ldots, i_{n_1}): 0 \leq i_1 \leq \cdots \leq i_{n_1} \leq n_2\}$$

and

$$R_2 = \{(j_1, \ldots, j_{n_1 + n_2}): 0 \leq j_1 \leq \cdots \leq j_{n_1 + n_2} \leq n_3\}.$$

But for $k = 1$ it is known from (2.1) or (1.31) that (some minor simplification may be necessary)

$$g_1(a(0), a(1), \ldots, a(n))$$
$$= \sum_{r=0}^{n-1} (-1)^{n-r-1} \binom{a(r) + 1}{n - r} g_1(a(0), a(1), \ldots, a(r)). \quad (2.29)$$

Combining (2.28) and (2.29), we have

$$g_3(A(n_1, n_2, n_3))$$
$$= \sum_{R_1} \sum_{R_2} \sum_{s=0}^{n_1} \sum_{t=i_s}^{i_{s+1}} \sum_{\substack{u=j_{s+t} \\ (s,t,u) \neq (n_1, n_2, n_3)}}^{j_{s+t+1}} \left[(-1)^{n_1+n_2+n_3-s-t-u-1} \right.$$
$$\left. \times \binom{a(s, t, u) + 1}{n_1 + n_2 + n_3 - s - t - u} \right] (g_1(a(0, 0, 0), \ldots, a(s, t, u))).$$
$$(2.30)$$

Consider the summation $\sum_{R_2} \sum_u$ which is equivalent to

$$\sum_{u=0}^{n_3} \sum_{0 \le j_1 \le \ldots \le j_{s+t} \le u} \sum_{u \le j_{s+t+1} \le \ldots \le j_{n_1+n_2} \le n_3}.$$

Since the term under summation is independent of $j_{s+t+1}, \ldots, j_{n_1+n_2}$, the summation over R_2 and u becomes

$$\sum_{u=0}^{n_3} \sum_{0 \le j_1 \le \ldots \le j_{s+t} \le u} \left[(-1)^{n_1+n_2+n_3-s-t-u-1} \binom{a(s, t, u) + 1}{n_1 + n_2 + n_3 - s - t - u} \right.$$
$$\left. \times g_1(a(0, 0, 0), \ldots, a(s, t, u)) \right]$$
$$\times \left[\sum_{j_{n_1+n_2}=u}^{n_3} \cdots \sum_{j_{s+t+2}=u}^{j_{s+t+3}} \sum_{j_{s+t+1}=u}^{j_{s+t+2}} 1 \right].$$

But the last factor simplifies to

$$\binom{n_1 + n_2 + n_3 - s - t - u}{n_1 + n_2 - s - t}.$$

Similar simplification can be made for the summation over R_1 and t. On implementing these simplifications in (2.30) we get

$$\sum_{\substack{s=0 \\ (s,t,u) \neq (n_1, n_2, n_3)}}^{n_1} \sum_{t=0}^{n_2} \sum_{u=0}^{n_3} \left[(-1)^{n_1+n_2+n_3-s-t-u-1} \binom{a(s, t, u) + 1}{n_1 - s, n_2 - t, n_3 - u} \right]$$
$$\times \left[\sum_{0 \le i_1 \le \ldots \le i_s \le t} \sum_{0 \le j_1 \le \ldots \le j_{s+t} \le u} g_1(a(0, 0, 0), \ldots, a(s, t, u)) \right] \quad (2.31)$$

because

$$\begin{pmatrix} a(s, t, u) + 1 \\ n_1 + n_2 + n_3 - s - t - u \end{pmatrix} \begin{pmatrix} n_1 + n_2 + n_3 - s - t - u \\ n_1 + n_2 - s - t \end{pmatrix}$$

$$\times \begin{pmatrix} n_1 + n_2 - s - t \\ n_1 - s \end{pmatrix} = \begin{pmatrix} a(s, t, u) + 1 \\ n_1 - s, n_2 - t, n_3 - u \end{pmatrix}.$$

However,

$$\sum_{0 \le i_1 \le \ldots \le i_s \le t} \sum_{0 \le j_1 \le \ldots \le j_{s+t} \le u} g_1(a(0, 0, 0), \ldots, a(s, t, u)) = g_3(A(s, t, u)),$$

(2.32)

by using (2.28). Now we can readily check that (2.27) follows from (2.30)–(2.32) for $k = 3$. The general proof is similar, and this completes the proof.

Let

$$A\begin{pmatrix} \mathbf{m} \\ \mathbf{n} \end{pmatrix} = \{a(\mathbf{r}): m_i \le r_i \le n_i, i = 1, \ldots, k\}.$$

Let $g_k(A(\begin{smallmatrix} \mathbf{m} \\ \mathbf{n} \end{smallmatrix}))$ denote the paths from the origin to the point $(n_0, n_1 - m_1, \ldots, n_k - m_k)$ with the restriction $\{a(\mathbf{m} + \mathbf{r}): 0 \le r_i \le n_i - m_i, i = 1, \ldots, k\}$. In this notation (2.27) is equivalent to

$$\sum_{\mathbf{m} \le \mathbf{r} \le \mathbf{n}} (-1)^{(\mathbf{n} - \mathbf{r}) \cdot \mathbf{1}} \begin{pmatrix} a(\mathbf{r}) + 1 \\ \mathbf{n} - \mathbf{r} \end{pmatrix} g_k \left(A\begin{pmatrix} \mathbf{m} \\ \mathbf{r} \end{pmatrix} \right) = \delta_{\mathbf{m}}^{\mathbf{n}}. \qquad (2.32a)$$

We introduce an ordering α on vectors different from $\le$, in which $\mathbf{x}\alpha\mathbf{y}$ means that for at least one i, $x_i < y_i$, $i = 1, \ldots, k$. Thus it is possible that simultaneously $\mathbf{x}\alpha\mathbf{y}$ and $\mathbf{y}\alpha\mathbf{x}$.

Consider the set $\{\mathbf{r}: \mathbf{0} \le \mathbf{r} \le \mathbf{n}\}$ of k-vectors. Denote its cardinality by d. Obviously $d = \prod_{i=1}^{k} (n_i + 1)$. Let $\mathbf{u}_1, \ldots, \mathbf{u}_d$ be an arrangement of this set of vectors such that $\mathbf{0} = \mathbf{u}_1 \alpha \mathbf{u}_2 \alpha \cdots \alpha \mathbf{u}_d = \mathbf{n}$. There are several ways of ordering the set. Denote by M_k and G_k two upper triangular matrices of order d as follows:

$$G_k = (-1)^{(\mathbf{u}_i - \mathbf{u}_j) \cdot \mathbf{1}} g_k \left(A\begin{pmatrix} \mathbf{u}_i \\ \mathbf{u}_j \end{pmatrix} \right) \quad \text{and} \quad M_k = \begin{pmatrix} a(\mathbf{u}_i) + 1 \\ \mathbf{u}_j - \mathbf{u}_i \end{pmatrix}, \quad (2.33)$$

where the (i, j)th element of each matrix is given, $i, j = 1, \ldots, d$.

COROLLARY 1

$$G_k M_k = I. \qquad (2.34)$$

Proof

The (i, j)th element of $G_k M_k$ is given by

$$\sum_{t=1}^{d} (-1)^{(\mathbf{u}_i - \mathbf{u}_t) \cdot \mathbf{1}} g_k \left(A \binom{\mathbf{u}_i}{\mathbf{u}_t} \right) \binom{a(\mathbf{u}_t) + 1}{\mathbf{u}_j - \mathbf{u}_t}. \tag{2.35}$$

Observing that

$$\binom{a(\mathbf{u}_t) + 1}{\mathbf{u}_j - \mathbf{u}_t} = 0$$

unless $\mathbf{u}_t \leq \mathbf{u}_j$ and

$$g_k \left(A \binom{\mathbf{u}_i}{\mathbf{u}_t} \right) = 0$$

unless $\mathbf{u}_i \leq \mathbf{u}_t$ and using (2.32a) in (2.35), we get

$$\sum_{\mathbf{u}_i \leq \mathbf{u}_t \leq \mathbf{u}_j} (-1)^{(\mathbf{u}_i - \mathbf{u}_t) \cdot \mathbf{1}} g_k \left(A \binom{\mathbf{u}_i}{\mathbf{u}_t} \right) \binom{a(\mathbf{u}_t) + 1}{\mathbf{u}_j - \mathbf{u}_t} = \delta_{\mathbf{u}_i}^{\mathbf{u}_j}.$$

The proof is complete when we note that $\mathbf{u}_i = \mathbf{u}_j$ if and only if $i = j$. The next corollary gives an explicit expression for $g_k(A(\mathbf{n}))$.

COROLLARY 2

$$g_k(A(\mathbf{n})) = (-1)^{d - \sum_{i=1}^{k} n_i - 1} \left\| \binom{a(\mathbf{u}_i) + 1}{\mathbf{u}_{j+1} - \mathbf{u}_i} \right\|, \qquad i, j = 1, \ldots, d - 1. \tag{2.36}$$

Proof

Clearly $g_k(A(\mathbf{n})) = g_k(A\binom{0}{\mathbf{n}})$. From Corollary 1 we see that

$$(-1)^{\sum_{i=1}^{k} n_i} g_k \left(A \binom{0}{\mathbf{n}} \right)$$

is equal to the $(1, d)$th element of M_k^{-1}. Some minor simplification helps us to complete the proof.

Taking into account both restrictions, let $g_k(A(\mathbf{n}) | B(\mathbf{n}))$ be the number of paths with upper restriction $A(\mathbf{n})$ and lower restriction

$B(\mathbf{n})$. We know that $x(\mathbf{r}) \leq a(\mathbf{r})$. However the nondecreasing condition on $b(\mathbf{r})$ leads us to $x(\mathbf{r}) \geq b(r_1, \ldots, r_{i-1}, r_i + 1, r_{i+1}, \ldots, r_k)$ if the path has moved from the $(r_1, \ldots, r_{i-1}, r_i + 1, r_{i+1}, \ldots, r_k)$th level $(i = 1, \ldots, k)$ to the **r**th level. Also, $x(n_1, \ldots, n_{i-1}, n_i - 1, n_{i+1}, \ldots, n_k) \geq b(\mathbf{n})$ for every i. For $k = 1$ it is easy to see that $b(r + 1) \leq x(r) \leq a(r)$, $r = 0, 1, \ldots, n_1 - 1$.

Proceeding similarly as in the case of an upper restriction, we obtain the following generalization of (2.27).

THEOREM 5

$$\sum_{0 \leq \mathbf{r} \leq \mathbf{n}} (-1)^{(\mathbf{n}-\mathbf{r}) \cdot \mathbf{1}} \binom{a(\mathbf{r}) - b(\mathbf{n}) + 1}{\mathbf{n} - \mathbf{r}}_+ g_k(A(\mathbf{r}) | B(\mathbf{r})) = \delta_\mathbf{0}^\mathbf{n}. \qquad (2.37)$$

If we introduce triangular matrices analogous to G_k and M_k with both lower and upper restrictions, the orthogonality of (2.34) can be established which in turn leads to the expression for $g_k(A(\mathbf{n}) | B(\mathbf{n}))$.

COROLLARY 3

$$g_k(A(\mathbf{n}) | B(\mathbf{n})) = (-1)^{d - \Sigma_{i=1}^{k} n_i - 1} \left\| \binom{a(\mathbf{u}_i) - b(\mathbf{u}_{j+1}) + 1}{\mathbf{u}_{j+1} - \mathbf{u}_i}_+ \right\|. \qquad (2.37a)$$

One way of choosing $\mathbf{u}_1, \ldots, \mathbf{u}_d$ is to let

$$\mathbf{u}_{i_1 + (n_1 + 1)i_2 + (n_1 + 1)(n_2 + 1)i_3 + \cdots + (n_1 + 1)\cdots(n_{k-1} + 1)i_k} = (i_1, i_2, \ldots, i_k)$$

for $0 \leq i_r \leq n_r$, $r = 1, \ldots, k$. (When $k = 1$, the natural ordering of $(0, 1, \ldots, n)$ determines $\mathbf{u}_1, \ldots, \mathbf{u}_d$, viz., $u_1 = 0, u_2 = 1, \ldots, u_{n+1} = n$.) Using this ordering let us find the number of lattice paths from $(0, 0, 0)$ to $(5, 1, 2)$ not crossing the surfaces determined by the equations $x = 3.7y$ and $x = z^2 - 1$. Here the lower restriction $B(1, 2)$ on the path does not exist, which is equivalent to saying $b(i, j) = 0$ for $i = 0, 1$ and $j = 0, 1, 2$, whereas the upper restriction $A(1, 2)$ on the path is given by

$$A(1, 2) = \begin{pmatrix} a(0, 0) = 1 & a(0, 1) = 1 & a(0, 2) = 1 \\ a(1, 0) = 2 & a(1, 1) = 5 & a(1, 2) = 5 \end{pmatrix}.$$

We take $\mathbf{u}_1 = (0, 0), \mathbf{u}_2 = (1, 0), \mathbf{u}_3 = (0, 1), \mathbf{u}_4 = (1, 1), \mathbf{u}_5 = (0, 2)$, and $\mathbf{u}_6 = (1, 2)$. From (2.36) we have

$g_2(A(1, 2))$

$$= (-1)^{6-3-1} \det \left| \binom{a(\mathbf{u}_i) + 1}{\mathbf{u}_{j+1} - \mathbf{u}_i} \right|, \qquad i, j = 1, \ldots, 5,$$

$$=
\begin{vmatrix}
\binom{a(0,0)+1}{1,0} & \binom{a(0,0)+1}{0,1} & \binom{a(0,0)+1}{1,1} & \binom{a(0,0)+1}{0,2} & \binom{a(0,0)+1}{1,2} \\
1 & 0 & \binom{a(1,0)+1}{0,1} & 0 & \binom{a(1,0)+1}{0,2} \\
0 & 1 & \binom{a(0,1)+1}{1,0} & \binom{a(0,1)+1}{0,1} & \binom{a(0,1)+1}{1,1} \\
0 & 0 & 1 & 0 & \binom{a(1,1)+1}{0,1} \\
0 & 0 & 0 & 1 & \binom{a(0,2)+1}{1,0}
\end{vmatrix}$$

$$=
\begin{vmatrix}
2 & 2 & 2 & 1 & 0 \\
1 & 0 & 3 & 0 & 3 \\
0 & 1 & 2 & 2 & 2 \\
0 & 0 & 1 & 0 & 6 \\
0 & 0 & 0 & 1 & 2
\end{vmatrix}
= 44 \cdot$$

We observe that the explicit solution to the enumeration problem (viz., Corollaries 2 and 3) comes out smoothly because of orthogonality guaranteed by the recurrence relation (viz., Theorems 4 and 5). If the nondecreasing property on boundaries is relaxed, it may be possible to get a recurrence relation other than (2.37), in which case an explicit solution might be difficult to obtain. This has prompted us to consider the recurrence relation as the main result.

5. Types of Diagonal Steps and a Correspondence

In the preceding chapter diagonal steps which are parallel to the line $x = y$ have been dealt with. Now we may consider other types of diagonal steps, for example, those which are parallel to $x = \mu y$ (see [15]).

In order to introduce various types of diagonal steps in a precise way, the following set of definitions leading to the definition of a path is useful. The reader is reminded of the fact that in Section 4 we gave only a representation of higher dimensional paths without any diagonal steps.

DEFINITIONS

Let $\mathbf{x} = (x_1, \ldots, x_k)$ and $\mathbf{y} = (y_1, \ldots, y_k)$ be two k-dimensional vectors.

(a) A step from $\mathbf{x}$ to $\mathbf{y}$ in k-dimensional space $(k \geq 0)$ is the ordered pair $(\mathbf{x}, \mathbf{y})$ satisfying

  (i) x_i and y_i are integers and $y_i - x_i \geq 0$, $i = 1, \ldots, k$;
  (ii) $\sum_{i=1}^{k} (y_i - x_i) \geq 1$;
  (iii) no integer greater than one divides every element of

  $$(y_1 - x_1, \ldots, y_k - x_k).$$

(b) A step from $\mathbf{x}$ to $\mathbf{y}$ is of type $\partial(k) = (\partial_1, \ldots, \partial_k)$ if $y_i - x_i = \partial_i$, $i = 1, \ldots, k$. Types $\partial_1(k) = (\partial_{11}, \ldots, \partial_{1k})$ and $\partial_2(k) = (\partial_{21}, \ldots, \partial_{2k})$ are different if and only if $\partial_{1j} \neq \partial_{2j}$ for some j.

(c) A lattice path (or briefly a path) in k-dimensional space is a sequence of vectors such that any two consecutive vectors form a step.

(d) A path from $\mathbf{a}$ to $\mathbf{b}$ is a path $(\mathbf{a}, \ldots, \mathbf{b})$.

If we represent the vectors by points in k-dimensional space, we get the usual geometric lattice path where every pair of consecutive points corresponding to the sequence of vectors is joined by a line segment, forming a step in the path. Thus note that

  (i) definition (a) includes diagonal steps, besides the steps on various axes:
  (ii) a step of type $\partial(k)$ which has only the jth $(j = 1, \ldots, k)$ element equal to one and others equal to zero is a unit step on axis X_j; and
  (iii) if $\partial(k)$ has at least two positive elements, the step is a diagonal one.

Consider the set $L_d(k)$ of k-dimensional paths, each of which starts from the origin and has r_i $(r_i > 0)$ steps of type $\partial_i(k)$, $i = 1, \ldots, m$, for some finite m. At any stage of the sequence of vectors in a path we may not be allowed to take the remaining steps unrestrictedly. In such a situation we say that the paths are restricted by boundaries. To write the next definition on boundaries we use the abbreviations $\mathbf{x}(-j)$ and $f(\mathbf{y})$, respectively, to denote the vector $\mathbf{x}$, without the jth element, and the function $f(y_1, \ldots, y_n)$ of the variables $y_1, \ldots, y_n$.

DEFINITION

(e) A boundary given by $x_j - f(\mathbf{x}(-j)) = 0$ which is called an X_j-boundary $(j = 1, \ldots, k)$, is not touched (crossed) by a path, if and only if, for every vector $\mathbf{y}$ in the path $f(\mathbf{y}(-j)) - y_j > 0 \ (\geq 0)$.

At this point, some comments deserve our attention. A particular boundary may both be an X_i-boundary and an X_j-boundary, $i \neq j$. For instance, $x_1 + x_2 = x_0 + a$ is an X_1-boundary as well as an X_2-boundary for paths to (m, m, m), $m > a$. We may have several X-boundaries or none at all. However unrestricted paths may be thought of as restricted even with trivial boundaries. For example, every path in $L_d(k)$ leading to $(\sum_{i=1}^{m} r_i \partial_{i1}, \ldots, \sum_{i=1}^{m} r_i \partial_{ik})$ has $x_j = \sum_{i=1}^{m} r_i \partial_{ij} + 1$ as a trivial X_j-boundary. Furthermore, all X-boundaries (including trivial ones) restrict uniquely the diagonal steps and these restrictions may be considered as boundaries for respective diagonal steps.

It has been conjectured in Section 7 of Chapter 1 that paths with diagonal steps might in some way be related to higher dimensional paths without diagonal steps. The main aim of this section is to establish an important correspondence supporting the above statement.

Let $L_d^*(k)$ be the set of paths $L_d(k)$ which do not touch a given X_j-boundary, $j = 1, \ldots, k$, some of which might be trivial. For each path in $L_d(k)$ construct a path by associating with each step of type $\partial_i(k)$ a unit step on axis X_i, $i = 1, \ldots, m$. Since X-boundaries uniquely determine boundaries for diagonal steps, we have essentially what we may call $\partial(k)$-boundaries. Put the $\partial_i(k)$-boundary on the transformed axis X_i. Denote by $L^*(m)$ the set of m-dimensional paths to $(r_1, \ldots, r_m)$ without diagonal steps which do not touch the boundaries determined by $\partial(k)$-boundaries. Then clearly by our construction we have a $1:1$ correspondence between $L_d^*(k)$ and $L^*(m)$ which is stated in the following lemma (see [6]).

LEMMA 1

$$|L_d^*(k)| = |L^*(m)|. \tag{2.38}$$

If the lattice paths have diagonal steps in addition to the steps on all axes, then $m > k$. In that case counting paths with diagonal steps is equivalent to counting paths without diagonal steps in a given higher dimension.

Finally, we illustrate the use of (2.38) in a particular problem. In view of the above lemma, counting paths with diagonal steps in some other cases could be possible with the help of the previous section.

We want to find the number of paths from $(0, 0)$ to $(\alpha + \beta n, n)$, $n \geq 0$, $\alpha > 0$, that never touch the line $x = \beta y$ ($\beta \geq 0$ being an integer) and have r_t (>0) diagonal steps parallel to $x = ty$, $t = 1, \ldots, \mu$ ($1 \leq \mu \leq \beta$, $\sum_{i=1}^{\mu} r_i \leq n$). Evidently each path has $\alpha + \beta n - \sum_{i=1}^{\mu} ir_i$ x-steps and $n - \sum_{i=1}^{\mu} r_i$ y-steps. Using the lemma, it is claimed that the number is equal to the number of $(\mu + 2)$-dimensional paths without diagonal steps from the origin to

$$A = \left(\alpha + \beta n - \sum_{i=1}^{\mu} ir_i, n - \sum_{i=1}^{\mu} r_i, r_1, \ldots, r_\mu\right)$$

that never touch the hyperplane

$$x_1 = \sum_{i=0}^{\mu} (\beta - i)x_{i+2}. \tag{2.39}$$

But we need to check that (2.39) represents boundaries for all types of steps. Suppose a path completes α_1 x-steps, α_i ($i = 2, \ldots, \mu + 2$) steps that are parallel to $x = (i - 2)y$. Then it has reached

$$\left(\alpha_1 + \sum_{j=2}^{\mu+2} (j - 2)\alpha_j, \sum_{j=2}^{\mu+2} \alpha_j\right),$$

from which point the line $x = \beta y$ is

$$\frac{\alpha_1 - \sum_{j=2}^{\mu+2} (\beta - j + 2)\alpha_j}{\beta - i + 2} \tag{2.40}$$

units away along the step parallel to $x = (i - 2)y$. However, after transformation, the new path reaches the point $(\alpha_1, \ldots, \alpha_{\mu+2})$ from which the distance of the hyperplane (2.39) along axis X_i ($i = 2, \ldots, \mu + 2$) is given by (2.40). This completes the verification. It is known (see Exercise 10, Chapter 1) that the number of paths from the origin to A that do not touch the hyperplane (2.39) is

$$\frac{\alpha}{\alpha + (\beta + 1)n - \sum_{i=1}^{\mu} ir_i} \binom{\alpha + (\beta + 1)n - \sum_{i=1}^{\mu} ir_i}{n - \sum_{i=1}^{\mu} r_i, r_1, \ldots, r_\mu},$$

which is therefore the desired expression.

6. Concluding Remarks

Theorem 3, which is a generalization of Theorem 1, has a continuous analogue, the applications of which are equally important (to be discussed in Chapter 4). Assume that the s_{ij}s, b_is, and a_is are real numbers instead of integers. Denote by $|(\mathbf{b}, \mathbf{a}; r)^*|$ the measure of the set $\{[s_{ij}]\}$ of $n \times r$ matrices which satisfy (2.14).

THEOREM 6

$$|(\mathbf{b}, \mathbf{a}; r)^*| = \|C_{ij}^{(r)*}\|_{n \times n}, \tag{2.41}$$

where

$$C_{ij}^{(r)*} = \frac{(a_i - b_j)_+^{r+j-i}}{(r+j-i)!} \quad \text{or} \quad C_{ij}^{(r)*} = \frac{(a_{n-j+1} - b_{n-i+1})_+^{r+j-i}}{(r+j-i)!}$$

and $(x)_+ = \max(0, x)$.

The proof is exactly similar to that of Theorem 3, where the summation signs in (2.18) are replaced by integration signs (see [13, 14]). Discussions of special cases, generalizations, and properties are analogous and are left out.

Another observation of significance is worth noting. Besides the natural temptation to use a combinatorial argument for the above combinatorial problems, probabilistic methods in certain instances may be applied with success. A particular probabilistic approach is suggested in Kemperman's monograph [7], where one embeds a Markov chain with absorbing states and uses properties of the Markov chain and matrices to derive the probability corresponding to the paths under consideration. For example, when $\mathbf{b} = (0, \ldots, 0, 1, 2, \ldots, \mu)$ and $\mathbf{a} = (c, c+1, \ldots, c+v, c+v, \ldots, c+v)$ in Theorem 1, the number of paths, which has been obtained by the repeated reflection principle in Chapter 1 [see (1.7)], is also derived along with an alternative expression by this method ([7, Section 8]) for a special case. In fact the alternative expression is very old and happens to be obtained by finding the $(n+v+c)$th power of the transition probability matrix (see [20]). These two expressions lead to the identity

$$\sum_{k=-\infty}^{\infty} \left[\binom{n+c+v}{c+v+kd}_+ - \binom{n+c+v}{c+v+e+kd}_+ \right]$$

$$= \frac{2^{n+c+v+1}}{d} \sum_{j=1}^{d-1} \left(\cos \frac{\pi j}{d} \right)^{n+c+v} \sin \frac{\pi j e}{d} \sin \frac{\pi j(a-b-e)}{d}, \tag{2.42}$$

where $d = n + c - u + 2$ and $e = c + v - u + 1$. The method works for more general situations from the Markov chain point of view than that described here. However one is not sure whether or not a suitable Markov chain can be found that would fit into the combinatorial problems discussed in this chapter.

Exercises

1. Prove Theorem 1 by the method of inclusion and exclusion ([21, 23]).
2. Derive (1.32) and (1.33) (one-A.P. case and two-A.P. case) directly from (2.6) (see [11, 12, 25]).
3. Prove that

$$D_r = \sum (-1)^{r + \phi(r)} \prod_{j=1}^{\phi(r)} \binom{a_{c_1 + \cdots + c_j} + 1}{c_j},$$

where the summation is over all compositions of r, $(c_1, \ldots, c_{\phi(r)})$ is a particular $\phi(r)$-composition of r (relation (10) in [5]), and $a_i = T_i - i$.
Note: The above expression is equivalent to the determinantal form of D_r in (2.6).
[Hint: Use (1.31) and induction.]

4. For α an integer, $0 \le \alpha \le a_1 + 1$, prove that [12]

$$\sum_{i=0}^{n} (-1)^i \binom{\alpha}{i} |(a_{i+1}, a_{i+2}, \ldots, a_n)| = |(a_1 - \alpha, a_2 - \alpha, \ldots, a_n - \alpha)|.$$

5. Prove property B for a one-boundary case. (See Section 2 and [3].)
6. Give a direct proof of

$$\| c_{ij} \|_{n \times n} \text{ (of (2.1))} = \| d_{ij} \|_{n \times n} \text{ (of (2.10))}.$$

7. Let $b_i \le a_i$, $i = 1, \ldots, n$. Prove

$$\begin{aligned}
|(a_1, &\ldots, a_n)| \\
&= 1 \cdot |(a_1 - b_1 - 1, a_2 - b_1 - 1, \ldots, a_n - b_1 - 1)| \\
&\quad + |(b_1)| \cdot |(a_2 - b_2 - 1, a_3 - b_2 - 1, \ldots, a_n - b_2 - 1)| \\
&\quad + |(b_1, b_2)| \cdot |(a_3 - b_3 - 1, a_4 - b_3 - 1, \ldots, a_n - b_3 - 1)| \\
&\quad + \cdots + |(b_1, \ldots, b_{n-1})| \cdot |(a_n - b_n - 1)| \\
&\quad + |(b_1, \ldots, b_n)| \cdot 1.
\end{aligned}$$

Note: (2.9) is a special case.

8. Use conjugation and obtain a new expression for (2.10).

9. Prove (2.12).

10. Prove (2.13).

11. We say that a k-composition $(t_1, \ldots, t_k)$ of m $[r]$-dominates a k-composition $(t'_1, \ldots, t'_k)$ of n if and only if

$$\sum_{i=1}^{j} t_i \geq r \sum_{i=1}^{j} t'_i, \qquad j = 1, \ldots, k.$$

(See Section 6, Chapter 1.) Let $(m, n; r)_k$ denote the set of possibilities that a k-composition of m $[r]$-dominates a k-composition of n. Prove that

$$|(m, n; r)_k| = \begin{cases} 0 & \text{for} \quad m < rn, \\ \binom{m}{k-1}\binom{n-1}{k-1} - r\binom{m-1}{k-2}\binom{n}{k} & \text{for} \quad m \geq rn. \end{cases}$$

Notice that $|(m, n; r)_k|$ represents the number of lattice paths from $(0, 0)$ to (m, n), not crossing the line $x = ry$, such that each path has exactly k horizontal and k vertical components. Hence show that the number of paths from $(0, 0)$ to (m, n) not crossing the line $x = ry$ is

$$\frac{m - rn + 1}{m + n + 1}\binom{m + n + 1}{n}.$$

(See [16].)

12. Prove the expression in (2.25).

[Hint: Follow the argument in Theorem 1 and Theorem 3.]

13. (a) Prove the expression in (2.26). Interpret it as a multidimensional path counting problem.

[Hint: Sum over s_{1j}s, s_{2j}s, $\ldots$, s_{rj}s recursively by remembering the conditions.]

(b) In the same problem change (i) to

$$c_i \leq s_{i1} \leq \cdots \leq s_{in}, \qquad i = 1, \ldots, r, \quad 0 \leq c_1 \leq \cdots \leq c_n.$$

Then prove that for nonnegative d_is the number of matrices is the determinant

$$\|d_{ij}\|_{(r+n) \times (r+n)},$$

where

$$d_{ij} = \begin{cases} \dfrac{\left(\begin{array}{c} a_{n-j+1} - b_{n-i+1} + \sum_{k=1}^{r-1} d_k + r \\ r+j-i \end{array}\right)_+}{}, & \begin{array}{l} i = 1, \ldots, r+n, \\ j = 1, \ldots, n, \end{array} \\[2em] \dfrac{\left(\begin{array}{c} u_{j-n} - b_{n-i+1} + \sum_{k=1}^{j-n-1} d_k + j - n - 1 \\ j-i \end{array}\right)_+}{}, & \end{cases}$$

$$\begin{array}{l} i = 1, \ldots, r+n, \\ j = n+1, \ldots, r+n, \end{array}$$

$$b_k = b_1 \quad \text{for} \quad k \le 0 \quad \text{and} \quad u_j = \max\!\left(b_1 - \sum_{k=1}^{j-1} d_k, c_j\right)$$

$$\text{for} \quad j = 1, \ldots, r.$$

14. Prove Theorem 5 and hence prove Corollary 3.

15. Using Corollary 2, rederive Exercise 10 of Chapter 1.
[Hint: Multiply each row by suitable terms, starting from the second and adding to the first row. Use identity (6.14) to make every element except the last entry in the first row zero.]

16. In the last example under diagonal steps prove that [15] if $\alpha = 0$, the number of paths, that do not touch except at the end is

$$\frac{1}{(\beta+1)n - \sum_{i=1}^{\mu} ir_i - 1} \left(\begin{array}{c} (\beta+1)n - \sum_{i=1}^{\mu} ir_i - 1 \\ n - \sum_{i=1}^{\mu} r_i, r_1, \ldots, r_\mu \end{array}\right).$$

17. In Bizley's problem (see Section 5, Chapter 1) if diagonal steps are allowed, show that [15] expressions for ψ_k (1.27) and ϕ_k (Exercise 11, Chapter 1) remain the same except that the expression for F_j is is changed to

$$\sum \frac{1}{k(m+n) - \sum_{i=1}^{\mu} ir_i} \left(\begin{array}{c} k(m+n) - \sum_{i=1}^{\mu} ir_i \\ kn - \sum_{i=1}^{\mu} r_i, r_1, \ldots, r_\mu \end{array}\right),$$

where the summation is over

$$R = \left\{ (r_1, \ldots, r_\mu) : 0 \le \sum_{i=1}^{\mu} r_i \le kn, \, 0 \le \sum_{i=1}^{\mu} ir_i \le km, \right.$$

$$\left. 0 \le r_i \le \min(km, kn), \, i = 1, \ldots, \mu \right\}.$$

18. Prove Theorem 6.

19. Give a direct proof of (2.42) [20].

20. Show the connection between (2.15) and the standard Young tableaux.

References

1. Berge, C., *Principles of Combinatorics.* Academic Press, New York, 1971.
2. Berlekamp, E. R., A class of convolution codes, *Infor. and Control* **6** (1963), 1–13.
3. Carlitz, L., Roselle, D. P., and Scoville, R. A., Some remarks on ballot-type sequences of positive integers, *J. Combin. Theory* **11** (1971), 258–271.
4. Fray, R. D., and Roselle, D. P., On weighted lattice paths, *J. Combin. Theory* **14** (1973), 21–29.
5. Göbel, F., Some remarks on ballot problems, *Math. Centrum Amsterdam*, Report S 321a, 1964.
6. Handa, B. R. and Mohanty, S. G., Higher dimensional lattice paths with diagonal steps, *Discrete Math.* **15** (1976), 137–140.
7. Kemperman, J. H. B., *The passage problem for a stationary Markov Chain*, Univ. of Chicago Press, Chicago, Illinois, 1961.
8. Kreweras, G., Sur une classe de problemes de denombrement liés aú treillis des partitions des entiers, *Cahiers du Bur. Univ. de Rech., Oper.* **6** (1965), 5–105.
9. Kryscio, R. J., Computational and estimation procedures in multidimensional right-shift processes with applications to epidemic theory. Ph.D. thesis submitted to the State University of New York at Buffalo, 1971.
10. MacMahon, P. A., *Combinatory Analysis*, Vols. I and II. Chelsea, Bronx, New York, 1960.
11. Mohanty, S. G., Some properties of compositions and their application to the ballot problem, *Canad. Math. Bull.* **8** (1965), 359–372.
12. Mohanty, S. G., Restricted compositions, *Fibonacci Quart.* **5** (1967), 223–234.
13. Mohanty, S. G., A short proof of Steck's result on two-sample Smirnov statistics, *Ann. Math. Statist.* **42** (1971), 413–414.
14. Mohanty, S. G., On more combinatorial methods in the theory of queues, Mathematical methods in queueing theory, *Lecture Notes in Economics and Mathematical Systems* **98**. Springer-Verlag, Berlin and New York, 1974.
15. Mohanty, S. G., and Handa, B. R., On lattice paths with several diagonal steps, *Canad. Math. Bull* **11** (1968), 537–545.
16. Mohanty, S. G., and Narayana, T. V., Some properties of compositions and their application to probability and statistics, I, *Biometrische Z.* **3** (1961), 252–258.
17. Narayana, T. V., Sur les treillis formés par les partitions d'un entier et leurs applications à la théorie des probabilitiés, *C.R. Acad. Sci. Paris* **240** (1955), 1188–1189.
18. Narayana, T. V., A combinatorial problem and its application to probability theory, 1, *J. Indian Soc. Agric. Statist.* **7** (1955), 169–178.
19. Narayana, T. V., A partial order and its applications to probability theory, *Sankhyā* **21** (1959), 91–98.
20. Percus, O. E. and Percus, J. K., Random walk and the comparison of two empirical distributions, *SIAM J. Appl. Math.* **18** (1970), 884–886.

21. Pitman, E. J. G., Simple proofs of Steck's determinantal expressions for probabilities in the Kolmogorov and Smirnov tests, *Bull. Austral. Math. Soc.* **7** (1972), 227–232.
22. Riordan, J., *Combinatorial Identities*. Wiley, New York, 1968.
23. Sarkadi, K., On the exact distributions of statistics of Kolmogorov–Smirnov type, *Period. Math. Hungar.* **3** (1973), 9–12.
24. Steck, G. P., The Smirnov two-sample tests as rank tests, *Ann. Math. Statist.* **40** (1969), 1449–1466.
25. Steck, G. P., Evaluation of some Steck determinants with applications, *Comm. Statist.* **3** (1974), 121–138.
26. Switzer, P., Significance probability bounds for rank orderings, *Ann. Math. Statist.* **35** (1964), 891–894.

3 Invariance and Fluctuation

1. Introduction

The content of this chapter differs somewhat from the main theme of previous chapters, viz., in counting lattice paths under some restrictions. In the study of fluctuations of certain random sequences combinatorial methods have been found to be extremely effective, and as such it is likely that the enumeration of paths will be encountered. Curiously enough, some particular sequences are such that the resulting paths, when cyclically or simply permuted, exhibit certain properties which are invariant with respect to the sequence selected. These properties lead to profound probabilistic results in fluctuation theory. We concentrate mainly on the study of some invariance properties and their applications in fluctuation theory. However, our interest and emphasis lie in lattice paths, and therefore the counting of paths is not completely neglected but is a result of our study.

In Chapter 1 a particular invariance property of cyclically permuted paths was used for counting paths under the line $x = \mu y$ by the method of penetrating analysis. With the help of the same property a more general problem, called Takács' urn problem (see Exercise 7 of Chapter 1), is solved. In Section 2 we reintroduce the urn problem and a precise formulation of the penetrating analysis. Besides a natural generalization of the urn problem, a different type of generalization, leading to certain combinatorial identities, is discussed. A further investigation of the invariance properties of cyclically permuted paths discussed in Section 3 gives rise to a refinement of a theorem by Chung and Feller. Other Chung–Feller theorems are also analyzed. Lastly, a well-known equivalence principle of Sparre Andersen in fluctuation theory is reexamined in the context of permutations of a given set of lattice paths, yielding a refinement of the principle.

2. Takács' Urn Problem

Takács' urn problem [30] is stated in Chapter 1 (see the exercises). The problem follows:

An urn contains n cards marked with nonnegative integers $k_1, \ldots, k_n$, respectively, such that $\sum_{i=1}^{n} k_i = k \leq n$. Cards are drawn without replacement from the urn. Let $X_r, r = 1, \ldots, n$, be the random variable representing the number on the card drawn at the rth drawing. Assume that all possible draws are equally likely. Then

$$P(X_1 + \cdots + X_r < r \quad \text{for} \quad r = 1, \ldots, n) = 1 - \frac{k}{n},$$

where $P(\cdot)$ denotes the probability of the event $(\cdot)$. When $k_1 = \cdots = k_a = 0$, $k_{a+1} = \cdots = k_n = \mu + 1$, the problem reduces to the ballot problem.

The solution follows directly from the next theorem, a combinatorial theorem on an invariance property of cyclic permutations of a set of nonnegative integers.

THEOREM 1

Suppose $k_1, \ldots, k_n$ are nonnegative integers so that $\sum_{i=1}^{n} k_i = k \leq n$. Among the n cyclic permutations of $(k_1, \ldots, k_n)$ there are exactly $n - k$ permutations for which the sum of the first r elements is less than r for every $r = 1, \ldots, n$.

Before providing the proof, certain remarks are in order. The vector $(a_1, \ldots, a_n)$, where $a_r = \sum_{i=1}^{r} k_i$, obviously represents a path from the origin to (k, n), beginning with a vertical unit or, by duality, represents the conjugate path (see Chapter 2, Section 2, for the definition) to (n, k), beginning with a horizontal unit. Interestingly, if $(u_1, \ldots, u_k)$ is the path conjugate to $(a_1, \ldots, a_n)$, the paths conjugate to the cyclic permutations of paths corresponding to $(a_1, \ldots, a_n)$ are represented by $U_i = (u_1^*, \ldots, u_k^*)$, $i = 0, 1, \ldots, n - 1$, where the elements are the nondecreasing sequence of $u_1 - i, u_2 - i, \ldots, u_k - i$ when reduced modulo (n). The theorem is proved by the method of penetrating analysis of Chapter 1, which is given a precise form by Takács in his book ([30, Chap. 1]). In order to gain a pictorial insight similar to that of Chapter 1, Section 4, one might take the reverse conjugate path into consideration.

Proof

Set

$$k_{n+r} = k_r, \quad a_r = k_1 + \cdots + k_r, \quad r = 1, 2, \ldots, \quad \text{and} \quad a_0 = 0.$$

Let

$$v_r = \begin{cases} 1 & \text{if} \quad i - r > a_i - a_r \quad \text{for every} \quad i > r, \\ 0 & \text{otherwise}, \end{cases} \tag{3.1}$$

and

$$w_r = \inf_{i \geq r} \{i - a_i\} \tag{3.2}$$

for $r = 0, 1, 2, \ldots$. It is evident that $\sum_{r=1}^{n} v_r$ represents the number of cyclic permutations of $(k_1, \ldots, k_n)$ with the desired property of the theorem. Moreover, $v_r = 0$ if and only if $w_r \neq r - a_r$, which is equivalent to $w_r = w_{r+1}$. Also, $v_r = 1$ if and only if $w_{r+1} - w_r = 1$. Combining these two statements, we get $v_r = w_{r+1} - w_r$. Since $a_{n+r} = a_n + a_r$, we have for $r = 0, 1, 2, \ldots$

$$v_{n+r} = v_r$$

and

$$w_{n+r} = \inf_{i \geq n+r} \{i - a_i\} = \inf_{i \geq r} \{n + i - a_{n+i}\} = n - k + w_r.$$

Thus

$$\sum_{r=1}^{n} v_r = w_{n+1} - w_1 = n - k, \tag{3.3}$$

which proves the theorem.

In the urn problem suppose there are c_j cards marked with j, $j = 0, 1, \ldots, k$. Then

$$\sum_{j=0}^{k} c_j = n \quad \text{and} \quad \sum_{j=0}^{k} jc_j = k.$$

Represent a sequence of cards drawn by a $(k + 1)$-dimensional path from the origin to $(c_0, c_1, \ldots, c_k)$ by associating a unit step on the X_j-axis to the number j on the drawn cards $j = 0, 1, \ldots, k$. It can be checked that the event $(X_1 + \cdots + X_r < r$ for $r = 1, \ldots, n)$ is represented by paths not touching the hyperplane $x_0 = \sum_{j=1}^{k} (j - 1)x_j$. Therefore from the solution of the urn problem we obtain the number of such paths equal to

$$\frac{n - k}{n} \binom{n}{c_1, \ldots, c_k} = \frac{\alpha}{\alpha + \sum_{j=1}^{k} jc_j} \binom{\alpha + \sum_{j=1}^{k} jc_j}{c_1, \ldots, c_k}, \qquad (3.4)$$

where

$$\alpha = c_0 - \sum_{j=1}^{k} (j - 1)c_j > 0.$$

Consider those js $(j \neq 0)$ for which $c_j > 0$; let these be $\mu_1 + 1, \ldots,$ $\mu_r + 1$, $r \leq k$; and let $n_0 = c_0$ and $n_i = c_{\mu_i + 1}$, $i = 1, \ldots, r$. Then an alternative form of (3.4) states that the number of paths from the origin to $(n_0, \ldots, n_r)$ that do not touch the hyperplane $x_0 = \sum_{i=1}^{r} \mu_i x_i$ is given by

$$\frac{\alpha}{\alpha + \sum_{i=1}^{r} (\mu_i + 1)n_i} \binom{\alpha + \sum_{i=1}^{r} (\mu_i + 1)n_i}{n_1, \ldots, n_r}, \qquad (3.5)$$

where $\alpha = n_0 - \sum_{i=1}^{r} \mu_i n_i$ and the $\mu_i s$ (≥ 0) are all different. Expression (3.5) is familiar to us from Chapter 1 (see Exercise 10).

As an application of Theorem 1 in probability theory we give below a proposition ([30, Chap. 2, Theorem 1]) on cyclically exchangeable random variables. $X_1, \ldots, X_n$ are called cyclically exchangeable random variables if for any $(x_1, \ldots, x_n)$ the joint distribution function of $X_1, \ldots, X_n$ at $(x_1, \ldots, x_n)$ is the same for all cyclic permutations of $(x_1, \ldots, x_n)$.

Let $X_1, \ldots, X_n$ be cyclically exchangeable random variables, each of which takes a nonnegative integer value. Then

$$P\left(\sum_{i=1}^{r} X_i < r \quad \text{for} \quad r = 1, \ldots, n \,\middle|\, \sum_{i=1}^{n} X_i = k\right)$$

$$= \begin{cases} 1 - \dfrac{k}{n} & \text{when} \quad 0 \leq k \leq n, \\ 0 & \text{otherwise,} \end{cases} \tag{3.6}$$

provided that the probability is defined. A simple corollary of (3.6) which finds applications in the theory of queues with batches is [23]

$$P\left(\sum_{i=1}^{r} X_i < \left[\frac{r-1}{m}\right] + 1 \quad \text{for} \quad r = 1, \ldots, n \,\middle|\, \sum_{i=1}^{n} X_i = k\right)$$

$$= \begin{cases} 1 - \dfrac{mk}{n} & \text{for} \quad 0 \leq mk \leq n, \\ 0 & \text{otherwise.} \end{cases} \tag{3.7}$$

An earlier proof of (3.6) by Takács [29] is rather probabilistic. For an alternative proof see [16].

The above result forms the core of Takács' book [30], which deals essentially with the application of combinatorial methods to fluctuation theory.

To end this section we point out another direction of generalization, emanating from (3.2) and (3.3), which leads to a set of combinatorial identities. For this purpose an identity equivalent to $w_{n+1} - w_1 = n - k$ is derived by examining the definition in (3.2). Since

$$dn + r + i - a_{dn+r+i} = d(n - k) + r + i - a_{r+i} \geq r + i - a_{r+i}$$

for $0 \leq i \leq n - 1$ and $d \geq 1$ an integer, we have

$$\inf_{i \geq r}\{i - a_i\} = \min\{r - a_r, r + 1 - a_{r+1}, \ldots, n + r - a_{n+r}\}$$

and

$$\inf_{i \geq r+1}\{i - a_i\} = \min\{r + 1 - a_{r+1}, r + 2 - a_{r+2}, \ldots, n + r - a_{n+r}\}.$$

But some elementary simplification shows that

$$\min\{n + r - a_{n+r}, n + r - 1 - a_{n+r-1}, \ldots, r - a_r\}$$

$$= \max\{0, 1 - k_r, 2 - k_r - k_{r-1}, \ldots, r - a_r, r + 1 - a_r - k_n,$$

$$r + 1 - a_r - k_n - k_{n-1}, \ldots, n - a_n\} + n + r - a_{n+r}$$

$$= \max\left\{0, x_{n+1-r}, x_{n+1-r} + x_{n+2-r}, \ldots, \sum_{i=1}^{r} x_{n+i-r},\right.$$

$$\left. \sum_{i=1}^{r} x_{n+i-r} + x_1, \sum_{i=1}^{r} x_{n+i-r} + x_1 + x_2, \ldots, \sum_{i=1}^{n} x_i \right\}$$

$$+ n + r - a_{n+r}, \tag{3.8}$$

where $x_i = 1 - k_{n+1-i}$. Let

$$\sigma = (x_1, \ldots, x_n), \quad s_j = \sum_{i=1}^{j} x_i, \quad M_j(\sigma) = \max\{s_1, \ldots, s_j\}, \quad 1 \le j \le n,$$

$$x^+ = \max\{0, x\}$$

and

$$\sigma_i = (x_i, x_{i+1}, \ldots, x_{n+i-1})$$

with $x_{n+r} = x_r, i = 1, \ldots, n$. We call σ_i the ith cyclic permutation of σ. Then because of (3.8) we get

$$w_{n+1} - w_1 = \sum_{i=1}^{n} (M_n^+(\sigma_i) - M_{n-1}^+(\sigma_i)).$$

Therefore from (3.3) it follows that

$$\sum_{i=1}^{n} (M_n^+(\sigma_i) - M_{n-1}^+(\sigma_i)) = n - k = s_n, \tag{3.9}$$

where $(x_1, \ldots, x_n)$ is a sequence of n numbers selected from $\{1, 0, -1, -2, \ldots\}$. Indeed, the restriction on the xs to be integers can be relaxed. Letting $(x_1, \ldots, x_n)$ be a sequence of n real numbers, the combinatorial identity

$$\sum_{i=1}^{n} (M_n^+(\sigma_i) - M_{n-1}^+(\sigma_i)) = s_n^+ \tag{3.10}$$

is proved in [8] and is generalized [14] to be

$$\sum_{i=1}^{n} (M_{r,n}^{+}(\sigma_i) - M_{r,n-1}^{+}(\sigma_i)) = s_n^{+}, \qquad (3.11)$$

where $M_{r,j}(\sigma)$ denotes the rth largest among $\{s_1, \ldots, s_j\}$, $1 \leq r \leq j \leq n$. For further generalized identities the reader may see [15, 21].

3. Chung–Feller Theorems

In Chapter 1, Section 3, it was proved that the number of paths to (n, n) that do not cross the line $x = y$ is the Catalan number $(n + 1)^{-1}\binom{2n}{n}$. Actually, a stronger result on paths to (n, n), giving the number $(n + 1)^{-1}\binom{2n}{n}$, which is a theorem of Chung–Feller [2], is known. We say that a path with vector representation $(a_1, \ldots, a_n)$ has r exceedances, $r = 0, 1, \ldots, n$, if exactly r inequalities among $a_j \geq j$, $j = 1, \ldots, n$, are satisfied. The Chung–Feller theorem states that the number of paths to (n, n) which have exactly r exceedances is independent of $r(r = 0, 1, \ldots, n)$ and hence equals $(n + 1)^{-1}\binom{2n}{n}$.

As a continuation of the last section, we study another interesting invariance property of the cyclic permutation of paths that gives us a refinement of the Chung–Feller theorem given by Narayana [24]. Consider the sequence of nonnegative integers $K = (k_1, \ldots, k_{n+1})$ such that $\sum_{i=1}^{n+1} k_i = n$. Let

$$A(K) = (a_1, \ldots, a_{n+1}), \qquad \text{where} \quad a_j = \sum_{i=1}^{j} k_i, \quad j = 1, \ldots, n + 1,$$

$$(3.12)$$

$$S(K) = \sum_{i=1}^{n+1} a_i, \qquad (3.13)$$

and

$$E(K) = |\{j : a_j \geq j, j = 1, \ldots, n + 1\}|. \qquad (3.14)$$

As usual $A(K)$ represents the vector corresponding to a path to $(n, n + 1)$, starting with a vertical unit. Clearly, $S(K)$ and $E(K)$ represent the area under the path and the exceedance number of the path, respectively.

THEOREM 2

Let K_i be the ith cyclic permutation of K (i.e., $K_i = (k_i, k_{i+1}, \ldots, k_{n+i})$ with $k_{n+r+1} = k_r$).

(i)   There exists a permutation, say, $(i_1, \ldots, i_{n+1})$ of $n + 1$ such that

$$S(K_{i_1}) > \cdots > S(K_{i_{n+1}}),$$

and

(ii) $E(K_{i_1}) > \cdots > E(K_{i_{n+1}})$.

Proof

Set $a_r = k_1 + \cdots + k_r$, $r = 1, 2, \ldots$. Then $a_{n+r+1} = a_{n+1} + a_r$.

$$A(K_i) = (a_i - a_{i-1}, a_{i+1} - a_{i-1} - a_{i-1}, \ldots, a_{n+i} - a_{i-1}) \quad (3.15)$$

and

$$S(K_i) = S(K_1) + (i - 1)n - (n + 1)a_{i-1} \quad (3.16)$$

since $a_{n+1} = n$. Thus for $u < v$ and $u, v = 1, \ldots, n + 1$,

$$S(K_u) - S(K_v) = (n + 1)(a_{v-1} - a_{u-1}) - (v - u)n$$

$$= (n + 1)(v - u)\left(\frac{a_{v-1} - a_{u-1}}{v - u} - \frac{n}{n + 1}\right). \quad (3.17)$$

Since $v - u$ is an integer such that $0 < v - u < n + 1$ and $a_{v-1} - a_{u-1}$ is a nonnegative integer, we conclude $S(K_v) \neq S(K_u)$, which proves (i).

To prove (ii) it is sufficient to show that for $u \neq v$ $S(K_u) > S(K_v)$ implies $E(K_u) > E(K_v)$. We see that the statement is true for $n = 1, 2$. Proceeding inductively, we assume that it holds good for all integers up to n, and we have to prove the statement for $n + 1$.

Theorem 1 asserts that among $K_1, \ldots, K_{n+1}$ there is exactly one cyclic permutation with zero exceedances that without loss of generality is K_1, say. Then K_1 must have $k_1 = 0$ and $k_{n+1} > 0$.

Consider

$$K_1^* = (k_2, \ldots, k_n, k_{n+1} - 1) = (k_1^*, \ldots, k_n^*)$$

and its cyclic permutations. Noting from (3.15) that

$$A(K_i^*) = (a_i^* - a_{i-1}^*, \ldots, a_n^* - a_{i-1}^*, a_n^* - a_{i-1}^* + a_1^*, \ldots, a_n^*) \quad (3.18)$$

and

$$A(K_{i+1}) = (a_i^* - a_{i-1}^*, \ldots, a_{n-1}^* - a_{i-1}^*, a_n^* + 1 - a_{i-1}^*,$$
$$a_n^* + 1 - a_{i-1}^*, a_n^* + 1 - a_{i-1}^* + a_1^*,$$
$$a_n^* + 1 - a_{i-1}^* + a_2^*, \ldots, a_n^* + 1), \qquad (3.19)$$

we get

$$S(K_{i+1}) = S(K_i^*) + i + 1 + a_n^* - a_{i-1}^*. \qquad (3.20)$$

Since K_1 has zero exceedances, $a_i < i$ for $i \leq n + 1$, and hence $a_{i-1}^* \leq i - 1$ and $a_n^* - a_{i-1}^* \geq n - i$. From this, and by comparing (3.18) and (3.19), it follows that

$$E(K_{i+1}) = E(K_i^*) + 1. \qquad (3.21)$$

As in (3.17) we have for $u < v$ and $u, v = 1, \ldots, n$,

$$S(K_u^*) - S(K_v^*) = n(v - u)\left(\frac{a_{v-1}^* - a_{u-1}^*}{v - u} - \frac{n - 1}{n}\right). \qquad (3.22)$$

Therefore

$$S(K_u^*) - S(K_v^*) > 0 \qquad \text{if and only if} \qquad \frac{a_{v-1}^* - a_{u-1}^*}{v - u} > \frac{n - 1}{n}.$$

Since $a_{v-1}^* - a_{u-1}^*$ and $v - u$ are integers,

$$S(K_u^*) - S(K_v^*) > 0 \qquad \text{if and only if} \qquad \cdot \frac{a_{v-1}^* - a_{u-1}^*}{v - u} \geq 1, \qquad (3.23)$$

i.e., if and only if

$$\frac{a_{v-1}^* - a_{u-1}^*}{v - u} > \frac{n}{n + 1}.$$

But from (3.20) we get

$$S(K_{u+1}) - S(K_{v+1}) > 0 \qquad \text{if and only if} \qquad \frac{a_{v-1}^* - a_{u-1}^*}{v - u} > \frac{n}{n + 1}.$$
$$\qquad (3.24)$$

Finally, from (3.23) and (3.24) we observe that $S(K_u^*) - S(K_v^*)$ and $S(K_{u+1}) - S(K_{v+1})$ have the same sign for $u \neq v$.

By the induction hypothesis it is assumed that for $u \neq v$, $u, v = 1, \ldots, n$,

$$S(K_u^*) > S(K_v^*) \qquad \text{implies} \qquad E(K_u^*) > E(K_v^*)$$

and thus using (3.23), we readily conclude that

$$S(K_{u+1}) > S(K_{v+1}) \quad \text{implies} \quad E(K_{u+1}) > E(K_{v+1}).$$

Lastly, it is easy to check that $S(K_1) < S(K_j)$ and $0 = E(K_1) < E(K_j), j = 2, \ldots, n + 1$. Thus the assertion is proved and this completes the proof of (ii).

Let $B_r = \{K: E(K) = r\}$. From Theorem 2(ii) it is immediate that $|B_0| = |B_1| = \cdots = |B_n|$, and thus the Chung–Feller theorem follows when we drop the initial vertical step of each path in Theorem 2. For other proofs see [18, 25].

It would be an interesting exercise to rewrite the theorem in terms of conjugate paths by recalling the conjugate path equivalence of cyclic permutation of paths (discussed just after the statement of Theorem 1).

Let $X_1, X_2, \ldots$ be a sequence of independently and identically distributed Bernoulli random variables such that $P(X_i = 1) = (X_i = -1) = \frac{1}{2}$. Let $S_j = \sum_{i=1}^{j} X_i$ be the jth partial sum and

$$N_n^* = |\{j: \text{either } S_j > 0 \text{ or } S_j = 0 \text{ and } S_{j-1} > 0, j = 1, \ldots, n\}|.$$

The above form of the Chung–Feller theorem can be rewritten in the language of fluctuation theory as

$$P(N_{2n}^* = 2r \,|\, S_{2n} = 0) = \frac{1}{n + 1}, \tag{3.25}$$

which is indeed Theorem 2A in [2].

When $S_{2n} = 2m, m > 0$, the conditional probability is also obtained as Theorem 2 in [2], which is stated as

$$P(N_{2n}^* = 2r \,|\, S_{2n} = 2m)$$

$$= \frac{2m}{\binom{2n}{n-m}} \sum_{j=0}^{r-m} \frac{1}{2(m+j)} \binom{2(m+j)}{j}$$

$$\times \frac{1}{2(n-m-j)+1} \binom{2(n-m-j)+1}{n-m-j}. \tag{3.26}$$

An important remark is in order. The Chung–Feller theorem is derivable as a special case from the following Spitzer combinatorial lemma [28]:

Let $\sigma = (x_1, \ldots, x_n)$ be a vector of real numbers such that $x_1 + \cdots + x_n = 0$ and no other partial sum of distinct elements vanishes. Then for each $r = 0, 1, \ldots, n - 1$ there is exactly one cyclic permutation of σ such that exactly r of its successive partial sums are positive.

Also, the theorem follows from Sparre Andersen's results in [27]. In either case an invariance property of cyclic permutations is utilized in all these results. Because of the invariance, such a combinatorial theorem can be converted into an analogous probabilistic theorem on cyclically exchangeable random variables. In addition to our earlier example in Section 2, we provide one more. In Chapter 8 of Takács' book [30] a special case of Spitzer's lemma is given as Theorem 1, an immediate probabilistic consequence of which is Theorem 2. We state it for completeness and observe that it is a generalization of (3.25).

Let $X_1, \ldots, X_n$ be cyclically exchangeable random variables that take on integer values. Denote by δ_n the number of positive partial sums among $X_1 + \cdots + X_r, r = 1, \ldots, n$. Then

$$P\left(\delta_n = j \,\middle|\, \sum_{i=1}^{n} X_i = 1\right) = \frac{1}{n} \qquad (3.27)$$

for $j = 1, \ldots, n$, provided that the conditional probability is defined.

We say that $X_1, \ldots, X_n$ are exchangeable random variables if for any $(x_1, \ldots, x_n)$ the joint distribution function of $X_1, \ldots, X_n$ at $(x_1, \ldots, x_n)$ is the same for all permutations of $(x_1, \ldots, x_n)$. Let $X_1, \ldots, X_n$ be exchangeable random variables taking on nonnegative integer values. Let

$$\Delta_n = |\{j : S_j < j, j = 1, \ldots, n\}|.$$

A combination of (3.6) and (3.27) leads to the expression of $P(\Delta_n = j | S_n = k)$ for various values of j and k (see [30, Section 37]).

In the ballot problem of Chapter 1 denote by α_r and β_r the number of votes for A and B, respectively, among the first r votes. Subscript r is called a strong lead position if $\alpha_r > \mu\beta_r$ is satisfied. Let $P_j(m, n)$ be the probability that there are j strong lead positions. The conditional distribution of Δ_n helps us to evaluate $P_j(m, n)$. Similarly, a weak lead position is defined as the subscript r for which $\alpha_r \geq \mu\beta_r$ is satisfied, and its distribution is obtained analogously [9]. Concerning fluctuation of election returns under Bizley's assumptions (Chapter 1, Section 5), see Takács [31], where the generating function technique is used.

We present a lemma to be used in the proof of the next theorem, which is another important result of Chung–Feller [2], giving the unconditional distribution of N_{2n}^* and leading to the well-known arcsine law in fluctuation theory.

LEMMA 1

$$L(n, n) \quad \Leftrightarrow \quad \bigcup_{j=0}^{n} L(2n - j, j: -1).$$

(See Chapter 1 for notation.)

Proof

Any path in $L(n, n)$ touches the line $x = y + k$ for some k, $k = 0$, $1, \ldots, n$; let $(s + k, s)$ be the last point where it touches. Reverse the segment of the path from $(0, 0)$ to $(s + k, s)$ and join with it the conjugate of the segment from $(s + k, s)$ to (n, n). The new path so constructed clearly belongs to $L(n + k, n - k: -1)$ so that it does not touch the line $x = y + k$ after $(s + k, s)$. The inverse mapping is obvious and unique. Thus the mapping is bijective and the lemma is proved.

THEOREM 3

$$P(N_{2n}^* = 2r) = \frac{1}{2^{2n}} \binom{2r}{r} \binom{2n - 2r}{n - r}. \tag{3.28}$$

Proof

Represent a $+1$ and a -1 by a horizontal unit and a vertical unit, respectively. Then the event $(N_{2n}^* = 2r)$ represents paths having $2n$ steps, of which $2r$ steps are below the line $x = y$. To prove the theorem it suffices to show that there exists a bijective mapping from the above set of paths to the set of paths to (n, n) passing through (r, r). Denote the two sets of paths by $A(n, r)$ and $B(n, r)$, respectively. The mapping of a path in $A(n, r)$ to a path in $B(n, r)$ consists of transforming the steps below the line $x = y$ to a segment from $(0, 0)$ to (r, r) and the steps above the line to a segment from (r, r) to (n, n). Let us partition $A(n, r)$ into four mutually exclusive cases:

(a) the first step is a horizontal one and the last is below the line $x = y$;

(b) the first step is a vertical one and the last is below the line $x = y$;

(c)   the first step is a horizontal one and the last is above the line $x = y$;

(d)   the first step is a vertical one and the last is above the line $x = y$.

We demonstrate the construction for case (a). Let a path belonging to this case cross the line $x = y$ at $(\alpha_1, \alpha_1), \ldots, (\alpha_{2t}, \alpha_{2t})$, $\alpha_1 < \alpha_2 < \cdots < \alpha_{2t}$. Clearly, the portion of the path from $(\alpha_{2t}, \alpha_{2t})$ to the end does not cross the line $x = y$ and has $2n - 2\alpha_{2t} \ (\leq 2r)$ steps. By the construction of Lemma 1 we first transform this to a path from $(r - n + \alpha_{2t}, r - n + \alpha_{2t})$ to (r, r), which forms a part of the new path. Next, for $j = 1, \ldots, t$ and $\alpha_0 = 0$ the segment of the new path from

$$\left(r - n + \alpha_{2t} - \sum_{i=1}^{j} (\alpha_{2t-2i+1} - \alpha_{2t-2i}), \right.$$

$$\left. r - n + \alpha_{2t} - \sum_{i=1}^{j} (\alpha_{2t-2i+1} - \alpha_{2t-2i}) \right)$$

to

$$\left(r - n + \alpha_{2t} - \sum_{i=1}^{j-1} (\alpha_{2t-2i+1} - \alpha_{2t-2i}), \right.$$

$$\left. r - n + \alpha_{2t} - \sum_{i=1}^{j-1} (\alpha_{2t-2i+1} - \alpha_{2t-2i}) \right) = C \quad \text{(say)}$$

is taken as the segment of the old path from

$$(\alpha_{2t-2j}, \alpha_{2t-2j}) \quad \text{to} \quad (\alpha_{2t-2j+1}, \alpha_{2t-2j+1})$$

or its conjugate, so that the new path crosses the line $x = y$ at C. In order to obtain the crossing we start from $j = 1$ and successively proceed to larger j.

Similarly, the segments above the line $x = y$ are joined one after the other, where the conjugate of the alternate segment is taken in order to obtain a crossing between two consecutive ones. The first segment or its conjugation is taken so that the first step after (r, r) becomes the same as the first step at the origin. It can easily be checked that the resulting path belongs to $B(n, r)$.

In other cases similar constructions can be performed in such a way that the first step after (r, r) and the step at the origin are different in cases (b) and (d) and are the same in case (c).

For the inverse mapping the following remarks play a vital role. The number of crossings of the path transformed from case (a) (case (d)) with the line $x = y$ before the point (r, r) is more (less) than the number of those after the point. The same number for a path transformed from case (b) (case (c)) is at least (at most) as large before (r, r) than after it. With the help of these remarks we should be able to go back from $B(n, r)$ to $A(n, r)$ uniquely. Thus the bijectivity of the mapping is established. The proof of the theorem is complete when we observe that

$$|B(n, r)| = \binom{2r}{r}\binom{2n - 2r}{n - r}.$$

The above proof by construction was given by Csáki and Vincze [4] whereas the original proofs of (3.25), (3.26), and (3.28) were based on the recurrence relation and the generating function method.

Using the Stirling approximation to (3.28), we get

$$\lim_{n \to \infty} P(N_n^* \leq \alpha n) = \frac{2}{\pi} \arcsin \alpha^{1/2}, \qquad 0 < \alpha < 1, \qquad (3.29)$$

which is the so-called arcsine law. In this sense Theorem 3 is called the finite arcsine law. The limiting form (3.29) is contained in a result of Erdös and Kac [10]. An analogue to Theorem 3 for exchangeable symmetric random variables is proved by Sparre Andersen [26], which, however, does not hold for arbitrary discrete variables. His proof again depends on an invariance property. The finite arcsine law of Sparre Andersen is stated as follows (also see [1, 17]):

Let $X_1, \ldots, X_n$ be exchangeable and symmetric random variables with continuous joint distribution. Define

$$N_n = |\{j: S_j > 0, j = 1, \ldots, n\}|.$$

Then

$$P(N_n = r) = \frac{1}{2^{2n}}\binom{2r}{r}\binom{2n - 2r}{n - r}. \qquad (3.30)$$

Comparison of (3.28) with (3.30) brings forth the striking similarity between the two results, which may be exploited further to derive other results. An illustration is as follows.

The sequence $S_0, S_1, \ldots, S_n$ of partial sums with $S_0 = 0$ is said to have a ladder index I if $S_r < S_I$ for $0 \leq r < I$. Let $p(n, j, k)$ be the probability that there is a kth ladder index (say, I_k) and that exactly j partial sums are larger than the kth ladder sum S_{I_k}. It can be proved [19, 20] by the correspondence between (3.28) and (3.30) that

$$p(n, j, k) = \frac{1}{2^{2n-k}} \binom{2j}{j} \binom{2n - 2j - k}{n - j}, \qquad j, k > 0, \quad j + k < n. \quad (3.31)$$

4. Sparre Andersen's Equivalence

Sparre Andersen [27] proves an interesting equivalence in fluctuation theory, relating the number of positive partial sums, the index for the first maximum, and the index for the last minimum of the partial sums, defined over a set of exchangeable random variables. The equivalence relation provides a method of obtaining an expression for the unconditional distribution of Δ_n of Section 3 (see [30, Section 37]). The infinite sequence of variables $X_1, X_2, \ldots$ are said to be exchangeable if $X_1, \ldots, X_n$ are exchangeable for $n = 2, 3, \ldots$. In addition to N_n define the events

$$L_n = \min\{j; S_j = \max\{S_0, S_1, \ldots, S_n\}\},$$

$$M_n = \max\{j: S_j = \min\{S_0, S_1, \ldots, S_n\}\}.$$

Sparre Andersen's equivalence relation can be stated as follows (see also [1, 12]):

Let $X_1, X_2, \ldots$ be exchangeable random variables and let C_n be an event defined on $X_1, \ldots, X_n$ which is invariant under permutations of $x_1, \ldots, x_n$. Then

$$P((L_n = k) \cap C_n) = P((M_n = n - k) \cap C_n) = P((N_n = k) \cap C_n).$$

$$(3.32)$$

The first equality is easily established by taking the reverse sequence $X_n, X_{n-1}, \ldots, X_1$ into consideration and remembering that the sequence is exchangeable. The relation (3.32) is true for independently and indentically distributed (i.i.d.) random variables since they are exchangeable.

Considering the sequence of i.i.d. Bernoulli variables, where

$$P(X = 1) = p \qquad \text{and} \qquad P(X = -1) = 1 - p, \qquad (3.33)$$

we give a refinement of the equivalence relation due to Csáki and Vincze [5] (where C_n is the event $S_n = 0$) which is derived by a simple construction based on the permutation of lattice paths. For this purpose, in addition to the events L_n, M_n, N_n, we define

$$R_n = |\{j: S_{j-1} = 0, S_j = 1, j = 1, \ldots, n\}|$$

and

$$T_n = \max\{S_0, S_1, \ldots, S_n\}.$$

THEOREM 4

Let $X_1, X_2, \ldots$ be a sequence of i.i.d. Bernoulli variables satisfying (3.33). Then

$$P(L_{2n-1} = k, T_{2n-1} = r \mid S_{2n} = 0) = P(N_{2n-1} = k, R_{2n-1} = r \mid S_{2n} = 0).$$

$$(3.34)$$

Proof

Obviously $P(S_m = 0) = 0$ if $m \neq 2n$, and therefore this case is omitted. If we represent a $+1$ by a horizontal unit and a -1 by a vertical unit, the event $S_{2n} = 0$ is represented by the paths from the origin to (n, n). Moreover, the event on the left-hand side of the equality corresponds to those paths to (n, n) which touch the line $x = y + r$ for the first time at the kth step, whereas the paths corresponding to the event on the right-hand side are those having k steps below the line $x = y$, of which r horizontal steps touch the line $x = y$.

To prove (3.34), we use a mapping from one set of paths to the other, which happens to be bijective. A typical path in the set of the right-hand side has r segments below the line $x = y$, where each segment is a path from (α, α) to $(\alpha + \beta, \alpha + \beta)$ that does not touch the line $x = y$ except at the endpoints. Let the sequence $(x_1, \ldots, x_{2n})$ of $+1$s and -1s correspond to this path. Then the uth segment is characterized by

$$(x_{i_u}, x_{i_u + 1}, \ldots, x_{i_u + j_u})$$

such that

$$s_{i_u - 1} = 0, \quad s_{i_u} > 0, \quad s_{i_u + 1} > 0, \quad \ldots, \quad s_{i_u + j_u} > 0, \quad s_{i_u + j_u + 1} = 0,$$
$$u = 1, \ldots, r, \quad \sum_{u=1}^{r} j_u = k.$$

From $(x_1, \ldots, x_{2n})$, construct the following new sequence of $+1$s and -1s as a permutation of $(x_1, \ldots, x_{2n})$:

$$
\begin{aligned}
(x_{i_1+j_1}, \quad & x_{i_1+j_1-1}, \ldots, x_{i_1}, \\
x_{i_2+j_2}, \quad & x_{i_2+j_2-1}, \ldots, x_{i_2}, \\
& \vdots \\
x_{i_r+j_r}, \quad & x_{i_r+j_r-1}, \ldots, x_{i_r}, \\
x_1, \quad & x_2, \qquad \ldots, x_{i_1-1}, \\
x_{i_1+j_1+1}, & x_{i_1+j_1+2}, \ldots, x_{i_2-1}, \\
& \vdots \\
x_{i_r+j_r+1}, & x_{i_r+j_r+2}, \ldots, x_{2n}).
\end{aligned}
$$

Observe that in this construction we have taken the reverse of certain segments of the path. This construction is illustrated in Fig. 1.

The new path clearly belongs to the set corresponding to the event on the left side. The inverse mapping is constructed by proceeding in a backward direction, and the bijectivity can be checked without any difficulty. The idea of the ladder index is used in the inverse mapping. This completes the proof.

Sparre Andersen's equivalence for this special case follows from (3.34). However it is not always possible to prove (3.32), in general, by the above elementary construction. In addition to the equivalence he has derived in [27] the exact distributions of L_{2n-1} under the condition $S_{2n} = 0$ for i.i.d. Bernoulli variables, which will be dealt with in the next chapter.

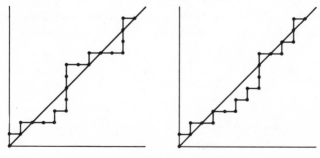

Figure 1

5. Concluding Remarks

Similarly to Chapter 2 we give a continuous analogue of Theorem 1. In (3.3) we demonstrated an invariance property of paths such that $w_{n+1} - w_1 = n - k$ no matter which sequence $k_1, \ldots, k_n$ is chosen so long as $\sum_{i=1}^{n} k_i = k$. Interestingly, this property is not restricted to only cyclic permutations of integers. It is true for more general situations (as suggested in the proof), as stated below without any proof ([30, Section 2]).

THEOREM 5

Suppose $u(x), 0 \leq x \leq t$, is a nondecreasing step function satisfying the condition $u(0) = 0$ and is defined over $t < x < \infty$ by the relation $u(t + y) = u(t) + u(y), y \geq 0$, and let

$$v(x) = \begin{cases} 1 & \text{if } y - x \geq u(y) - u(x) \quad \text{for } y \geq x, \\ 0 & \text{otherwise.} \end{cases} \tag{3.35}$$

Then

$$\int_0^t v(x)\, dx = \begin{cases} t - u(t) & \text{when } u(t) \leq t, \\ 0 & \text{otherwise.} \end{cases} \tag{3.36}$$

Its probabilistic interpretation is given in terms of stochastic processes with cyclically exchangeable increments in [30, Section 13]. A closely related result by the same author [29] is as follows:

Let $X_1, \ldots, X_n$ be nonnegative exchangeable random variables and let $Y_1, \ldots, Y_n$ be the order statistics from uniform distribution over the interval $(0, t)$. Let $\{X_i\}$ and $\{Y_i\}$ be independent of each other. Then

$$P\left(\sum_{i=1}^{r} X_i \leq Y_r \,\middle|\, \sum_{i=1}^{n} X_i = u \right) = \begin{cases} 1 - \dfrac{u}{t} & \text{if } 0 \leq u \leq t, \\ 0 & \text{otherwise.} \end{cases} \tag{3.37}$$

An alternative proof, which is based on (2.41), is provided in [22].

There are similar variations and generalizations of (3.36) (see [3, 6, 13]).

Besides Sparre Andersen's equivalence, one may find other equivalence relations (see Feller [11, Chap. 3]) in the case $p = \frac{1}{2}$ in (3.33) such as,

$$P(S_i \geq 0 \quad \text{for} \quad i = 1, \ldots, 2n) = 2P(S_i > 0 \quad \text{for} \quad i = 1, \ldots, 2n)$$

$$(3.38)$$

and

$$P(S_{2n} = 0) = P(S_i \neq 0 \quad \text{for} \quad i = 1, \ldots, 2n). \quad (3.39)$$

Construction procedures similar to those used in Theorem 4 are applied in [7] to prove (3.38) and (3.39).

In conclusion, a word of caution to the reader would be helpful. There is no pretense that the study of invariance and its application to fluctuation theory is in any way complete. We content ourselves to focus our attention only on some areas that are related to lattice paths.

Exercises

1. Prove (3.6) and (3.7).
2. Check (3.8).
3. Let σ_j^* be the jth reverse cyclic permutation $(x_j, x_{j-1}, \ldots, x_1, x_n, x_{n-1}, \ldots, x_{j+1})$, $m_{r,k}(\sigma)$ be the rth smallest among $\{s_1, \ldots, s_k\}$, and $x^- = \min\{0, x\}$. Then prove that [15]

$$\sum_{i=1}^{n} (M_{r,k}^+(\sigma_i) + m_{r,k}^-(\sigma_i^*)) = (k - r + 1)s_n,$$

and show that (3.11) is a special case.
4. Referring to the remark following the statement of Theorem 1, prove that if $(u_1, \ldots, u_k)$ is the path conjugate to $(a_1, \ldots, a_n)$, the paths conjugate to the cyclic permutations of paths corresponding to $(a_1, \ldots, a_n)$ are represented by $U_i = (u_1^*, \ldots, u_k^*)$, $i = 0, 1, \ldots, n - 1$, where the elements are the nondecreasing sequence of $u_1 - i, u_2 - i, \ldots, u_k - i$ when reduced modulo (n). Also show that Theorem 2(i) follows from the above.
5. Prove $S(K_1) < S(K_j)$ and $E(K_1) < E(K_j), j = 2, \ldots, n + 1$.
6. Prove (3.26).
7. Prove Spitzer's combinatorial lemma and show that Chung–Feller theorem (3.25) follows from the lemma. Also show that (3.27) is a special case of the lemma.

8. (a) Find expressions for $P(\Delta_n = j \,|\, S_n = k)$ for various values of j and k.
    (b) Find an expression for $P_j(m, n)$.
    (c) Let $Q_j(m, n)$ represent the probability that there are j weak lead positions. Find an expression for $Q_j(m, n)$.

9. Prove (3.30).

10. Prove (3.31).

11. Prove (3.32).

12. Prove (3.37).

13. Prove the Csáki–Tusnády formulation which is as follows: Suppose that there are n points $P_1, \ldots, P_n$ on a directed circle of unit circumference and that there is a positive number q such that $nq < 1$. If Q is an arbitrary point on the circle, put the points $Q_1, \ldots, Q_n$ on the circle in the positive direction from the point Q such that the length of arc QQ_k is kq. We say that Q is a point of first category if the arc QQ_k contains less than k of the points $P_1, \ldots, P_n$ for $k = 1, \ldots, n$. Then the measure of the set of points of first category is $1 - nq$.

14. Prove (3.38) and (3.39), using a construction procedure.

References

1. Baxter, G., Combinatorial methods in fluctuation theory, *Z. Wahrsch. Verw. Gebiete* **1** (1963), 263–270.

2. Chung, K. L., and Feller, W., On fluctuations in coin tossing, *Proc. Nat. Acad. Sci. U.S.A.* **35** (1949), 605–608.

3. Csáki, E., and Tusnády, G., On the number of intersections and the ballot theorem, *Period. Math. Hungar.* **2** (1972), 5–13.

4. Csáki, E., and Vincze, I., On some distributions connected with the arcsine law, *Publ. Math. Inst. Hungar. Acad. Sci.* **8** (1963), 281–291.

5. Csáki, E., and Vincze, I., On some combinatorial relations concerning the symmetric random walk, *Acta Sci. Math.* **24** (1963), 231–235.

6. Daniels, H. E., The statistical theory of the strength of bundles of threads I, *Proc. Roy. Soc. London Ser. A* **183** (1945), 405–597.

7. Doherty, M., An amusing proof in fluctuation theory, Combinatorial mathematics III, *Lecture Notes in Mathematics* **452**, Springer-Verlag, Berlin and New York, 1975.

8. Dwass, M., A fluctuation theorem for cyclic random variables, *Ann. Math. Statist.* **33** (1962), 1450–1454.

9. Engelberg, O., Generalizations of the ballot problem, *Z. Wahrsch. Verw. Gebiete* **3** (1965), 271–275.

10. Erdős, P., and Kac, M., On the number of positive sums of independent random variables, *Bull. Amer. Math. Soc.* **53** (1947), 1011–1020.

11. Feller, W., *An Introduction to Probability Theory and Its Applications*, Vol. 1, 2nd ed. Wiley, New York, 1957.

12. Feller, W., On combinatorial methods in fluctuation theory, in *Probability and Statistics, The Herald Cramér Volume.* (V. Gremander, ed.), pp. 75–91. Almqvist and Wiksell, Stockholm and Wiley, New York, 1959.

13. Gehér, L., On a theorem of L. Takács, *Act. Sci. Math.* **29** (1968), 163–165.

14. Graham, R. L., A combinatorial theorem for partial sums, *Ann. Math. Statist.* **34** (1963), 1600–1602.

15. Harper, L. H., A family of combinatorial identities, *Ann. Math. Statist.* **37** (1966), 509–512.

16. Heyde, C. C., A derivation of the ballot theorem from the Spitzer–Pollaczek identity, *Proc. Cambridge Philos. Soc.* **65** (1969), 755–757.

17. Hobby, C., and Pyke, R., Combinatorial results in fluctuation theory, *Ann. Math. Statist.* **34** (1963), 1233–1242.

18. Hodges, J. L., Galton's rank test, *Biometrika* **42** (1952), 261–262.

19. Imhof, J. P., On ladder indices and random walk, *Z. Wahrsch. Verw. Gebiete* **9** (1967), 10–15.

20. Imhof, J. P., Some joint laws in fluctuation theory, *Ann. Math. Statist.* **42** (1971), 1099–1103.

21. Mohanty, S. G., A note on combinatorial identities for partial sums, *Canad. Math. Bull.* **14** (1971), 65–67.

22. Mohanty, S. G., On more combinatorial methods in the theory of queues, Mathematical methods in queueing theory, *Lecture Notes in Economics and Mathematical Systems* **98**, Springer-Verlag, Berlin and New York, 1974.

23. Mohanty, S. G., and Jain, J. L., On the two types of queueing process involving batches, *Canad. Oper. Res. Soc. J.* **8** (1970), 38–43.

24. Narayana, T. V., Cyclic permutation of lattice paths and the Chung–Feller theorem, *Skand. Aktuarietidskr.* (1967), 23–30.

25. Sarkady, K., On Galton's rank order test, *Publ. Math. Inst. Hungar. Acad. Sci.* **7** (1962), 127–131.

26. Sparre Andersen, E., On the number of positive sums of random variable, *Skand. Aktuarietidskr.* **32** (1949), 27–36.

27. Sparre Andersen, E., On sums of symmetrically dependent random variables, *Skand. Aktuarietidskr.* **36** (1953), 123–138.

28. Spitzer, F., A combinatorial lemma and its application to probability theory, *Trans. Amer. Math. Soc.* **82** (1956), 323–339.

29. Takács, L., Combinatorial methods in the theory of queues, *Rev. Internat. Statist. Inst.* **32** (1964), 207–219.

30. Takács, L., *Combinatorial Methods in the Theory of Stochastic Processes.* Wiley, New York, 1967.

31. Takács, L., On the fluctuations of election returns, *J. Appl. Probab.* **7** (1970), 114–123.

4 Random Walk and Rank Order Statistics

1. Introduction

The present chapter deals with applications of path counting to distributional problems in random walks and statistical nonparametric inferences. Discrete random walks by their very nature, especially when the destination at a given time is fixed, can be represented by paths, and thus any probability distribution of characteristics defined on a random walk involves the enumeration of paths with restrictions constrained by the characteristics. In Section 2 we obtain several distributions and joint distributions based on earlier results. An excellent reference on random walks is Spitzer's book [53], which, however, differs substantially from the present material in its treatment and content. Because of a very interesting property, rank order statistics arising in nonparametric inferences are directly related to paths. Using this relation (which we call Gnedenko's technique), problems of the distribution of these statistics are handled in Section 3. In this chapter, unlike the earlier ones, the interest often goes beyond the earlier restrictions, viz., boundaries. The restrictions might be, for example, on the number of crossings of a particular line by the paths. Because of the correspondence with paths, one may observe that a particular distributional problem can be formulated in terms of either a restricted random walk or rank order statistics. In some instances a proof in this section consists of a combination of the direct counting results of Chapters 1 and 2 and the constructional procedures of Chapter 3. An alternative but unified approach due to Dwass is discussed in the next section, in which paths play an indirect role via random walks. A comparison between the two techniques is made. Our approach in this chapter is to illustrate the applications of the use of various techniques but not to list all possible derivations of distributions. In conclusion, besides other remarks we bring out the continuous counterpart of rank order statistics, which as observed earlier cannot be related to lattice paths. However the basic treatment and tools remain the same for these cases as well. The reader may be cautioned mildly that previous exposure to statistics would be helpful in going through the material of this chapter.

2. Random Walks

It is well known that random walk models serve as a first approximation to the theory of Brownian motion. Random walks have also helped in finding the distribution of mainly rank order statistics that arise in two-sample problems.

A particle is said to perform a random walk on a line in the usual sense when, starting from an initial position on the line (usually an integer), it moves at any stage either a unit step $(+1)$ in the positive direction with probability p $(0 < p < 1)$ or a unit step (-1) in the negative direction with probability $1 - p$. Let this be called the simple random walk. The random walk could have either absorbing or reflecting barriers on one or both sides. Often the interest lies in the position of the particle at any given stage, and in this sense the random walk is considered to be restricted.

Random walk problems have been extensively discussed in [16], where a few common problems have been dealt with thoroughly by various methods. For example, the distribution of the first passage time (i.e., the particle at a certain stage reaches a given point for the first time) has been obtained by the elementary combinatorial method, the generating function method, the method of difference equations, and the general method of Markov chains, of which the first three are essentially combinatorial in nature. It is instructive to recall Feller's remark in this regard that sometimes an elementary combinatorial argument would enable us to solve a technically difficult problem by replacing a formidable analysis apparatus. With this spirit in mind and in conjunction with the results on lattice paths, we examine a few problems of interest in random walks.

To illustrate the point, let us for simplicity derive the probability that a random walk starting from the origin has the first passage through k at the $(2n + k)$th stage. Representing each $+1$ by a horizontal unit, each -1 by a vertical unit, and following Feller's argument [16, Chap. 3, p. 73], we see that the random walk is represented by paths from the origin to $(n + k, n)$, not touching the line $x = y + k$ except at the end such that each path has probability $p^n(1 - p)^{n-k}$. From Chapter 1 we know that the number of such paths is $[k/(2n + k)]\binom{2n+k}{n}$, and therefore the desired probability is

$$\frac{k}{2n + k} \binom{2n + k}{n} p^n(1 - p)^{n-k}. \tag{4.1}$$

Denoting by $R(m, n)$ the one-dimensional random walk which starts from the origin and reaches the point $m - n$ in $m + n$ steps, one may observe from this simple illustration that the evaluation of the distribution of any random variable defined over the walk needs only the enumeration of the paths from the origin to (m, n), with restrictions determined by the random variables, since each path has the same probability $p^m(1 - p)^n$. In our situation the restricted random walk may be represented by a sequence $(X_1, \ldots, X_{m+n})$ of independent Bernoulli random variables, where

$$X_i = \begin{cases} +1 & \text{with probability } p, \\ -1 & \text{with probability } 1 - p. \end{cases}$$

Equivalently, it is also represented by $(S_0, S_1, \ldots, S_{m+n})$, where $S_0 = 0$, $S_j = X_1 + \cdots + X_j$, $j = 1, \ldots, m + n$. We give a few more results related to the counting of paths.

In the sequel two identities that play a vital role are

$$\sum_{k=0}^{n} \frac{a}{a + \mu k} \binom{a + \mu k}{k} \frac{b}{b + \mu(n - k)} \binom{b + \mu(n - k)}{n - k}$$

$$= \frac{a + b}{a + b + \mu n} \binom{a + b + \mu n}{n} \tag{4.2}$$

and

$$\sum_{k=0}^{n} \binom{a + \mu k}{k} \frac{b}{b + \mu(n - k)} \binom{b + \mu(n - k)}{n - k} = \binom{a + b + \mu n}{n}. \quad (4.3)$$

When a, b, and μ are positive integers, (4.2) and (4.3) can easily be proved by consideration of paths, and in fact these are given as exercises in Chapter 1. However, the identities are true in general which can be seen in [18] (see Chapter 6). It is fun to see that because of their importance in applications these identities have appeared in the literature so often in various forms that it is hard to keep track of every single situation. We cite only one by Rohatgi [48]. These identities have been generalized in [39] to the similar ones involving multinomial coefficients. From the path point of view a general identity which includes (4.2) and (4.3) as special cases is immediate. (See Exercise 7, Chapter 2.) It is given by

$$
\begin{aligned}
|(a_1, &\ldots, a_n)| \\
&= 1 \cdot |(a_1 - b_1 - 1, a_2 - b_1 - 1, \ldots, a_n - b_1 - 1)| \\
&\quad + |(b_1)| \cdot |(a_2 - b_2 - 1, a_3 - b_2 - 1, \ldots, a_n - b_2 - 1)| \\
&\quad + |(b_1, b_2)| \cdot |(a_3 - b_3 - 1, a_4 - b_3 - 1, \ldots, a_n - b_3 - 1)| \\
&\quad + \cdots + |(b_1, \ldots, b_{n-1})| \cdot |(a_n - b_n - 1)| \\
&\quad + |(b_1, \ldots, n_n)| \cdot 1, \quad\quad\quad\quad\quad\quad\quad\quad\quad (4.4)
\end{aligned}
$$

where $|(a_1, \ldots, a_n)| = 0$ if either any $a_i < 0$ or $a_1 \leq a_2 \leq \cdots \leq a_n$ is not satisfied.

Let $N_r(m, n)$ be the number of cases in which the particle in $R(m, n)$ crosses the origin r times. In other words

$$N_r(m, n) = |\{(s_0, s_1, \ldots, s_{m+n}): s_{m+n} = m - n, \beta = r\}|,$$

where $\beta = |\{j: s_j = 0, s_{j-1} \cdot s_{j+1} = -1\}|$. Kanwar Sen [30] has derived the following theorem:

THEOREM 1

(a) $\quad N_r(n, n) = \dfrac{2(r + 1)}{n} \binom{2n}{n - r - 1}.$

(b) For $m > n$

$$N_r(m, n) = \frac{m - n + 2r + 1}{m + n + 1} \binom{m + n + 1}{n - r}.$$

Proof

$N_r(n, n)$ is equal to the number of paths from the origin to (n, n), each of which crosses the line $x = y$ r times. Let $N_r(n, n)_1 = N_r(n, n)$ when the first step is a horizontal one; $N_r(n, n)_2 = N_r(n, n)$ when the first step is a vertical one. Then

$$N_r(n, n) = N_r(n, n)_1 + N_r(n, n)_2$$

and

$$N_r(n, n)_1 = N_r(n, n)_2.$$

Therefore we have to show that

$$N_r(n, n)_1 = \frac{r + 1}{n} \binom{2n}{n - r - 1}.$$

Clearly it is so for $r = 0$, which is the well-known Catalan number.
Using induction,

$$N_r(n, n)_1 = \sum_{k=1}^{n-r} N_0(k, k)_1 N_{r-1}(n - k, n - k)_2$$

$$= \sum_{k=1}^{n-r} \frac{1}{k} \binom{2k}{k - 1} \frac{r}{n - k} \binom{2n - 2k}{n - r - k}$$

$$= \frac{r + 1}{n} \binom{2n}{n - r - 1} \qquad \text{(by (4.2))},$$

which completes the verification of (a).
For $m > n$

$$N_r(m, n) = \begin{cases} \displaystyle\sum_{k=r}^{n} N_{r-1}(k, k)_1 N_0(m - k, n - k) & \text{if } r \text{ is even,} \\[2em] \displaystyle\sum_{k=r}^{n} N_{r-1}(k, k)_2 N_0(m - k, n - k) & \text{if } r \text{ is odd.} \end{cases}$$

But $N_0(m - k, n - k)$ is equal to the number of paths from the origin to $(m - k, n - k)$ that do not cross the line $x = y$, and the expression is

$$\frac{m - n + 1}{m + n - 2k + 1} \binom{m + n - 2k + 1}{n - k}$$

from (1.5). Hence

$$N_r(m, n) = \sum_{k=r}^{n} \frac{r}{k} \binom{2k}{k - r} \frac{m - n + 1}{m + n - 2k + 1} \binom{m + n - 2k + 1}{n - k}$$

$$= \frac{m - n + 2r + 1}{m + n + 1} \binom{m + n + 1}{n - r}$$

by (4.2) as before. This completes the proof of (b).

In the above we have followed the line of proof in [40]. An alternative proof is possible by constructing a bijective mapping between the above set of paths and the set of paths from the origin to $(m + r, n - r)$ that does not cross the line $x = y$ (see [30]). This is left as an exercise.

Next we consider a general situation in which a boundary that is not to be crossed by any path is given by the path $C = (c_1, \ldots, c_n), c_n = m$ (see [40]). Let $(m - c_j, n - j + 1), j = 1, \ldots, n$, be called the jth node of C. We say that a path $A = (a_1, \ldots, a_n)$ touches C exactly at r ($r = 1, \ldots, s$) nodes among the first s nodes ($s = 1, \ldots, n$) of C when

(i) $a_{i_j} = c_{i_j}$ for any subset $(i_1, \ldots, i_r)$ of $(1, \ldots, s)$ and $a_i < c_i$ for others among $a_1, \ldots, a_s$, and

(ii) $a_i \leq c_i, i = s + 1, \ldots, n$.

Denote by $D(r, s; C)$ the number of such paths.

THEOREM 2

$$D(r, s; C) = \begin{cases} 1, & \text{when} \quad r = s = n; \\ |(c_{r+1} - c_r, \ldots, c_n - c_r)|, & \text{when} \quad r = s < n; \\ |(c_{r+1} - 1, \ldots, c_n - 1)|, & \text{when} \quad r < s = n; \\ |(c_{r+1} - 1, \ldots, c_{n-1} - 1, c_n)| - |(c_r - 1, \ldots, c_{n-1} - 1)|, \\ \qquad \text{when} \quad r < s = n - 1; \\ |(c_{r+1} - 1, \ldots, c_s - 1, c_{s+1}, \ldots, c_n)| \\ \quad - |(c_r - 1, \ldots, c_s - 1, c_{s+2}, \ldots, c_n)|, \\ \qquad \text{when} \quad r < s < n. \end{cases}$$

$$(4.5)$$

Proof

Trivially, the first two expressions are true. To prove the next we use induction. Clearly,

$$D(1, n; C) = 1 \cdot |(c_2 - c_1 - 1, \ldots, c_n - c_1 - 1)|$$

$$+ \sum_{i=1}^{n-2} |(c_1 - 1, \ldots, c_i - 1)|$$

$$\cdot |(c_{i+2} - c_{i+1} - 1, \ldots, c_n - c_{i+1} - 1)|$$

$$+ |(c_1 - 1, \ldots, c_{n-1} - 1)| \cdot 1.$$

Substituting $a_i = c_{i+1} - 1$, $b_i = c_i - 1$, $i = 1, \ldots, n - 1$, in (4.4), we obtain

$$D(1, n; c) = |(c_2 - 1, \ldots, c_n - 1)|.$$

By the induction hypothesis

$$D(r, n; C) = |(c_{r+1} - c_1 - 1, \ldots, c_n - c_1 - 1)|$$

$$+ \sum_{i=1}^{n-r-1} |(c_1 - 1, \ldots, c_i - 1)|$$

$$\cdot |(c_{i+r+1} - c_{i+1} - 1, \ldots, c_n - c_{i+1} - 1)|$$

$$+ |(c_1 - 1, \ldots, c_{n-r} - 1)|$$

$$= |(c_{r+1} - 1, \ldots, c_n - 1)| \qquad \text{(by (4.4))}.$$

To prove the rest, let j ($j = r, r + 1, \ldots, s$) be the last node where the path touches C. For $s < n - 1$ we can write

$$D(r, s; C) = 1 \cdot |(c_{r+1} - c_r - 1, \ldots, c_s - c_r - 1,$$

$$c_{s+1} - c_r, \ldots, c_n - c_r)|$$

$$+ \sum_{i=r}^{s-2} |(c_r - 1, \ldots, c_i - 1)|$$

$$\cdot |(c_{i+2} - c_{i+1} - 1, \ldots, c_s - c_{i+1} - 1,$$

$$c_{s+1} - c_{i+1}, \ldots, c_n - c_{i+1})|$$

$$+ |(c_r - 1, \ldots, c_{s-1} - 1)| \cdot |(c_{s+1} - c_s, \ldots, c_n - c_s)|.$$

$$(4.6)$$

The first factor represents the number of paths each of which touches C at exactly $r - 1$ nodes before it touches the last node at j, and its expression is obtained from the third part of the theorem. As before the application of (4.4) simplifies (4.6) to the desired expression. When $s = n - 1$, the proof is similar. This completes the proof.

Results similar to (4.5) may be established for any consecutive s nodes.

The interpretation of Theorem 2 in terms of a generalized random walk is interesting. In a random walk the particle starting from the origin moves at any stage either a unit in the positive direction or $c_{i+1} - c_i$ units $(i = 1, \ldots, n - 1)$ in the negative direction if the particle has moved $(i - 1)$ times to the left during stages preceding the present stage. Note that $c_{i+1} - c_i = 0$ implies that the particle stays at a given point, and this move is counted as a move in the negative direction. Representing each move in the positive direction by a horizontal unit and in the negative direction by a vertical unit, we can show that $D(r, n; C)$ is equal to the number of ways in which the particle reaches (never crosses) c_1 exactly r times in $m + n$ steps.

We state a corollary and leave the proof as an exercise.

COROLLARY

The number of paths from $(0, 0)$ to (m, n) that touch the line $x = \mu(y - t)$ exactly r (>0) times, where μ, t are integers, $\mu > 0$, and $m > \mu(n - t)$, is given by

$$\sum_{k=0}^{t-1} \frac{m - \mu(n - t - r)}{m + n - (\mu + 1)k - r + \mu t}$$

$$\times \binom{m + n - (\mu + 1)k - r + \mu t}{n - r - k}\binom{(\mu + 1)k - \mu t}{k}$$

$$- \sum_{k=0}^{t-2} \frac{m - \mu(n - t - r + 1)}{m + n - (\mu + 1)k - r + \mu(t - 1)}$$

$$\times \binom{m + n - (\mu + 1)k - r + \mu(t - 1)}{n - r - k}$$

$$\times \binom{(\mu + 1)k - \mu(t - 1)}{k} \qquad \text{if} \quad t > 0, \qquad (4.7)$$

$$\frac{m - \mu(n - r)}{m + n - r} \binom{m + n - r}{n - r} \qquad \text{if} \quad t = 0. \qquad (4.8)$$

When $\mu = 1$, (4.7) is reduced to

$$\frac{m - n + 2t + r - 1}{m + n - r + 1} \binom{m + n - r + 1}{n - t - r + 1}. \tag{4.9}$$

We may also be interested in finding the number of paths that touch the given boundary C at least r times; we mention only the particular case for which $c_s = c_{s+1} = \cdots = c_n$. In this situation

$$\sum_{k=r}^{s} D(k, s; C) = N(\underbrace{c_{r+1}, \ldots, c_s, c_s + 1, \ldots, c_s + 1}_{n-s})$$

$$- N(\underbrace{c_r, \ldots, c_s, c_s + 1, \ldots, c_s + 1}_{n-s-1}). \tag{4.10}$$

Consider the random walk denoted by $R(m, n; \mu)$, where the particle, initially at the origin, moves at any stage either $+1$ or $-\mu$ unit steps (μ being a positive integer) and reaches the point $m - \mu n$ in $m + n$ steps. Let χ be be the maximum value attained by the particle in $R(\mu n, n; \mu)$. From (4.10) we are able to find the distribution of χ. Denote by $N(\cdot)$, the number of cases in which $(\cdot)$ occurs. Then it can be proved [26] that

$$N(\chi = r) = \begin{cases} \dfrac{1}{(\mu + 1)n + 1} \dbinom{(\mu + 1)n + 1}{n} \\ \qquad \text{when} \quad r = 0; \\[2ex] \displaystyle\sum_{\alpha=0}^{n-s} \dfrac{r}{(\mu + 1)\alpha + r} \dbinom{(\mu + 1)\alpha + r}{\alpha} \\ \qquad \times \dbinom{(\mu + 1)(n - \alpha) - r}{n - \alpha} \\[2ex] \qquad - \displaystyle\sum_{k=s}^{n} \dfrac{r + 1}{(\mu + 1)(n - k) + r + 1} \\ \qquad \times \dbinom{(\mu + 1)(n - k) + r + 1}{n - k}\dbinom{(\mu + 1)k - r - 1}{k} \\[2ex] \qquad \text{when} \quad r = (s - 1)\mu + v, \quad s = 1, \ldots, n, \\ \qquad\qquad\qquad\qquad\qquad\qquad v = 1, \ldots, \mu; \\[2ex] 0 \qquad \text{otherwise.} \end{cases}$$

$$\tag{4.11}$$

If $r \leq \mu$, the expression simplifies to

$$\frac{\mu + r + 1}{(\mu + 1)n + r} \binom{(\mu + 1)n + r}{n - 1},$$ (4.12)

and when $\mu = 1$, we have

$$N(\chi = r) = \frac{2r + 1}{2n + 1} \binom{2n + 1}{n - r}.$$ (4.13)

As a final application of the above elementary combinatorial technique we obtain the joint distribution of two characteristics defined below [26]. In $R(\mu n, n; \mu)$ assume "condition d" viz., $S_i S_{i+1} \geq 0$ for all i, and let

$$\lambda = |\{j : S_{j-1} = 0, S_j = 1\}|,$$

and

$$\pi = |\{j : S_j > 0\}|.$$

Here λ and π represent the number of times the particle enters the positive side and the number of stages in the positive side, respectively. In terms of paths "condition d" means that paths are not allowed to cross the line $x = \mu y$ except through lattice points.

THEOREM 3

$$N(\lambda = k, \pi = r \mid d) = \begin{cases} \dfrac{1}{(\mu + 1)n + 1} \binom{(\mu + 1)n + 1}{n} \\ \qquad \text{when} \quad k = r = 0; \\[2mm] \left[\dfrac{\mu k}{(\mu + 1)(\alpha - k) + \mu k} \binom{(\mu + 1)(\alpha - k) + \mu k}{\alpha - k} \right] \\ \qquad \times \left[\dfrac{k + 1}{(\mu + 1)(n - \alpha) + k + 1} \right. \\ \qquad \times \left. \binom{(\mu + 1)(n - \alpha) + k + 1}{n - \alpha} \right] \\ \qquad \text{when} \quad r = (\mu + 1)\alpha - k; \\ \qquad\qquad \alpha = k, k + 1, \ldots, n; k = 1, \ldots, n; \\ 0 \qquad \text{otherwise.} \end{cases}$$ (4.14)

Proof

$N(\lambda = 0, \pi = 0 | d)$ represents the number of paths that are above the line $x = \mu y$ but do not cross it and therefore equals the first expression in (4.14). Moreover, π cannot take any other values except those in the theorem.

For $k = 1$ and $r = (\mu + 1)\alpha - 1$ a typical path of this kind, which is shown in Fig. 1, consists of two segments:

(i) the segment up to $(\mu t, t)$, $t = 0, 1, \ldots, n - \alpha$, that lies above $x = \mu y$ and never crosses it, and

(ii) the segment starting from $(\mu t, t)$ that initially contains a horizontal step, intersects the line $x = \mu y$ at $(\mu(t + \alpha), t + \alpha)$ without having touched it before, and finally ends at $(\mu n, n)$ without crossing $x = \mu y$ after $(\mu(t + \alpha), t + \alpha)$.

Therefore

$N(\lambda = 1, \pi = (\mu + 1)\alpha - 1 | d)$

$$
= \sum_{t=0}^{n-\alpha} \frac{\mu}{(\mu + 1)\alpha - 1} \binom{(\mu + 1)\alpha - 1}{\alpha - 1}
$$

$$
\times \frac{1}{(\mu + 1)t + 1} \binom{(\mu + 1)t + 1}{t} \frac{1}{(\mu + 1)(n - \alpha - t) + 1}
$$

$$
\times \binom{(\mu + 1)(n - \alpha - t) + 1}{n - \alpha - t}
$$

$$
= \frac{\mu}{(\mu + 1)\alpha - 1} \binom{(\mu + 1)\alpha - 1}{\alpha - 1} \frac{2}{(\mu + 1)(n - \alpha) + 2}
$$

$$
\times \binom{(\mu + 1)(n - \alpha) + 2}{n - \alpha}
$$

by using (4.2). This confirms the theorem for $k = 1$.

In general, any path to be counted for

$$
N(\lambda = k + 1, \pi = (\mu + 1)\alpha - k - 1 | d)
$$

consists of two segments of which the first one is up to $(\mu y, y)$ with $\lambda = k$ and

$$
\pi = (\mu + 1)(\alpha - \beta) - k \ (y = \alpha - \beta, \ldots, n - \alpha; \beta = 1, \ldots, \alpha - k).
$$

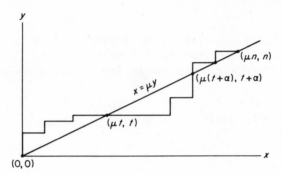

Figure 1

Using induction, we obtain

$$N(\lambda = k + 1, \pi = (\mu + 1)\alpha - k - 1 | d)$$

$$= \sum_{\beta=1}^{\alpha-1} \sum_{y=\alpha-\beta}^{n-\alpha} \frac{\mu k}{(\mu + 1)(\alpha - \beta - k) + \mu k}$$

$$\times \binom{(\mu + 1)(\alpha - \beta - k) + \mu k}{\alpha - \beta - k} \frac{k + 1}{(\mu + 1)(y - \alpha + \beta) + k + 1}$$

$$\times \binom{(\mu + 1)(y - \alpha + \beta) + k + 1}{y - \alpha + \beta}$$

$$\times \frac{\mu}{(\mu + 1)(\beta - 1) + \mu} \binom{(\mu + 1)(\beta - 1) + \mu}{\beta - 1}$$

$$\times \frac{1}{(\mu + 1)(n - y - \beta) + 1} \binom{(\mu + 1)(n - y - \beta) + 1}{n - y - \beta},$$

which with the help of (4.2) simplifies to the required expression of the theorem for $\lambda = k + 1$. This completes the proof.

From (4.14), we readily obtain

$$N(\lambda = k | d) = \begin{cases} \dfrac{(\mu + 1)k + 1}{(\mu + 1)n + 1} \dbinom{(\mu + 1)n + 1}{n - k}, & k = 0, 1, \ldots, n, \\ \\ 0 & \text{otherwise.} \quad (4.15) \end{cases}$$

Also, we can establish the relation

$$N(\chi = r) = \begin{cases} N(\lambda = r \,|\, d) & \text{when} \quad r = 0, \\[2mm] \displaystyle\sum_{i=0}^{r-1} \binom{r-1}{i} N(\lambda = r - i \,|\, d) & \text{when} \quad 1 \le r \le \mu. \end{cases}$$

(4.16)

Another relation of interest is

$$N(\lambda = k, \pi = r \,|\, d) = N(\psi = -k, \phi = r),$$ (4.17)

where

$$\psi = \min_{0 \le i \le (\mu+1)n} \{S_i\}$$

and

$$\phi = \min\{i : S_i = \psi, \text{ there exists } (i_0, i_1, \ldots, i_{-\psi})$$
$$\text{such that } 0 = i_0 < i_1 < \cdots < i_{-\psi} = i \text{ and}$$
$$S_{i_{j-1}+1} > -j, S_{i_{j-1}+2} > -j, \ldots, S_{i_j-1} > -j,$$
$$S_{i_j} = -j, j = 1, 2, \ldots, -\psi\}.$$

The proof of (4.17) can be done by finding the joint distribution of ψ and ϕ and verifying the equality (see [26]). Observe that when $\mu = 1$, (4.17) coincides with (3.34), which is derived by using a bijective mapping. For a generalization of results in [26] that is done in a routine manner see [42].

It is a common practice to apply the generating function method in solving combinatorial problems, and thus an illustration of the application of the technique perhaps will be adequate for our purpose. In fact the identities (4.2) and (4.3) can be proved by the use of the generating function. By employing both path and generating functions in $R(n, n)$ Jain [28] has obtained the joint distribution of (i) positive steps, (ii) returns to zero from the positive side, (iii) returns to zero from the negative side, and (iv) crossings of zero. To be more precise the following terminology and notations are needed:

(i) The ith step is said to be positive or negative if $S_i - S_{i-1} > 0$ or <0, respectively.
(ii) The random walk is said to have a positive (negative) wave from i to $i + k$ if $S_i = 0$, $S_{i+k} = 0$, and S_{i+j} has the positive (negative) sign for $j = 1, \ldots, k - 1$.

(iii) $\gamma = |\{i\colon S_i = 0,\ S_{i-1}S_{i+1} = -1\}|$; i.e., γ is the number of crossings of the origin.

(iv) $R(n; u, v; g, h)$: The random walk initially at the origin returns to the origin at the $2n$th step, which has u positive waves, v negative waves, $2g$ positive steps, and $\gamma = h$.

(v) $H(m, k)$: The random walk starting from the origin reaches k for the first time at the mth step.

THEOREM 4

$$N(R(n; u, v; g, h)) = \begin{cases} 2\binom{u-1}{t-1}\binom{v-1}{t-1}\alpha \\[2mm] \quad\text{when}\quad h = 2t - 1, \\[4mm] \dfrac{u + v - 2t}{t}\binom{u-1}{t-1}\binom{v-1}{t-1}\alpha \\[2mm] \quad\text{when}\quad h = 2t \neq 0, \end{cases} \qquad (4.18)$$

where

$$\alpha = \frac{u}{2g - u}\binom{2g - u}{g}\frac{v}{2n - 2g - v}\binom{2n - 2g - v}{n - g}.$$

Proof

We observe that in our problem, the number of cases to form u positive waves comprising $2g$ positive steps is denoted by $N(N_{2g-1} = 2g - u,\ R_{2g-1} = u | S_{2g} = 0)$ in the notation of Section 4 of Chapter 3, which is equal to $N(H(2g - u, u))$ when we use the construction of Theorem 4 of the Chapter 3. Thus

$$N(N_{2g-1} = 2g - u,\ R_{2g-1} = u | S_{2g} = 0) = \frac{u}{2g - u}\binom{2g - k}{g}. \qquad (4.19)$$

Similarly, the number of cases in which v negative waves consisting of $2n - 2g$ negative steps can be formed is given by

$$N(N_{2n-2g-1} = 2n - 2g - v,\ R_{2n-2g-1} = v | S_{2n-2g} = 0)$$

$$= \frac{v}{2n - 2g - v}\binom{2n - 2g - v}{n - g}. \qquad (4.20)$$

In order to use the generating function method, let us identify the exponent of x with the number of crossings and that of y with the number of negative waves. Assume that $u \geq v$. Then by inserting one or more negative waves between two consecutive positive waves, we get only two crossings and, of course, no crossings if we do not put in any negative wave. Moreover, there are $u - 1$ positions between u positive waves where negative waves can be inserted. Second, the negative waves can also be placed before the first or after the last positive wave, in which case only one crossing is formed. Combining all these facts, we write the bivariate generating function of crossings and negative waves as

$$(1 + x^2(y + y^2 + \cdots))^{u-1}(1 + x(y + y^2 + \cdots))^2$$

$$= \left(1 + \frac{x^2 y}{1 - y}\right)^{u-1} \left(1 + \frac{xy}{1 - y}\right)^2,$$

in which the coefficients of $x^{2t-1}y^v$ and $x^{2t}y^v$ can be checked to be

$$2\binom{u-1}{t-1}\binom{v-1}{t-1} \quad \text{and} \quad \frac{u+v-2t}{t}\binom{u-1}{t-1}\binom{v-1}{t-1}, \quad (4.21)$$

respectively. Thus for $u \geq v$ the theorem is proved from (4.19)–(4.21). It is also true for $u < v$ when we interchange the role of positive and negative waves.

Jain [28] further obtains the joint generating functions of all these variables directly from the definition by using the result of Theorem 4. The generating functions are utilized to derive various marginal distributions in a routine manner.

The treatment of random walks with barriers neither stands in isolation nor poses any particular problem. Indeed it has been indirectly covered earlier. For example, the random walk $R(m, n)$ with an absorbing barrier at k needs counting paths from $(0, 0)$ to (m, n) which do not touch the line $x = y + k$. We give Exercise 8 on absorbing barriers, in which one is required to obtain the corresponding generating function. On the other hand, suppose there is a partially reflecting barrier at k $(k > m - n \geq 0)$, so that as soon as the particle reaches k it either is reflected back to $k - 1$ with probability q or gets absorbed at k with probability $1 - q$. Denote by $P(m, n; r)$ the probability that the random walk has r reflections. The computation of $P(m, n; r)$ obviously

involves the evaluation of the number of paths from the origin to
(m, n) that touch the line $x = y + k$ exactly r times. Thus,

$$
P(m, n; r) = \begin{cases}
\left[\binom{m + n}{n} - \binom{m + n}{m - k} \right] p^m (1 - p) \\
\qquad \text{when} \quad r = 0, \\[2mm]
\dfrac{2k - m + n + r - 1}{m + n - r + 1} \binom{m + n - r + 1}{m - k - r + 1} \\[2mm]
\qquad \times \left(\dfrac{q}{1 - p} \right)^r p^m (1 - p)^n \\[2mm]
\qquad \text{when} \quad 1 \le r \le m - k + 1.
\end{cases}
\tag{4.22}
$$

The first expression follows from (1.2) and the second from (4.9).

Finally, we give a result by Handa [25], involving a partially reflecting barrier and the assumption that the particle at any stage is also
allowed to stay in its position. Let the particle at any stage move -1
unit, $+1$ unit, or stay in its position with corresponding probabilities
p_1, p_2, and p_0 ($p_0 + p_1 + p_2 = 1$). Other assumptions do not change.
In terms of paths no movement is equivalent to a diagonal step. Without any ambiguity we only state the next theorem in the language of
lattice paths. Denote by $P(m, n, t; r)$ the probability that a path with t
diagonal steps ends at (m, n) and has r reflections at the line $x = y + k$.

THEOREM 5

$$
P(m, n, t; 0) = \left[\binom{m + n - t}{t, n - t} - \binom{m + n - t}{t, m - t - k} \right] p_0^t p_1^{n - t} p_2^{m - t},
$$

$$
P(m, n, t; r) = \frac{2k - m + n + r - 1}{m + n - t - r + 1} \binom{m + n - t - r + 1}{t, m - t - k - r + 1}
$$

$$
\times \left(\frac{q}{p_1} \right)^r p_0^t p_1^{n - t} p_2^{m - t} \qquad \text{for} \quad 1 \le r \le m - t - k + 1.
$$

$$
\tag{4.23}
$$

The proof depends on counting paths with diagonal steps and on
combinatorial identities involving multinomial coefficients which arise
as the natural generalization of (4.2) and (4.3) (see [39]). Handa [25] has

also dealt with different types of diagonal steps (see Chapter 2) and multitype reflections at the reflection barrier (also see [22, 36]). Consideration of two barriers is similar.

3. Rank Order Statistics—Gnedenko's Technique

The rank order statistics are commonly used for nonparametric testing problems in the two-sample case. To make the idea clear let us consider two independent sets $X_1, \ldots, X_m$ and $Y_1, \ldots, Y_n$ of i.i.d. random variables having continuous distributions. Define the rank order indicator of $\{X_1, \ldots, X_m, Y_1, \ldots, Y_n\}$ as a vector $(Z_1, \ldots, Z_{m+n})$ such that Z_j takes the value $+1$ or -1 if the jth minimum among $\{X_1, \ldots, X_m, Y_1, \ldots, Y_n\}, j = 1, \ldots, m + n,$ is X_t for some $t \in \{1, \ldots, m\}$ or Y_t for some $t \in \{1, \ldots, n\}$, respectively. Obviously $(Z_1, \ldots, Z_{m+n})$ is a sequence of $m + 1$s and $n - 1$s. Any random variable defined on the rank order indicator $(Z_1, \ldots, Z_{m+n})$ is called a rank order statistic.

Run, median, rank sum, Kolmogorov–Smirnov (briefly K–S), and Galton statistics are some examples of rank order statistics.

$R_{m,n}$ (run statistic):

$R_{m,n} = 1 +$ number of changes either of type $(1, -1)$
    or type $(-1, 1)$ in the sequence $(Z_1, \ldots, Z_{m+n})$.

$M_{m,n}$ (median statistic):

$$M_{m,n} = \left| \left\{ j: j \le \left[\frac{m+n+1}{2} \right], Z_j = 1 \right\} \right|.$$

$U_{m,n}$ (rank sum statistic):

$$U_{m,n} = \sum_{i=1}^{m} \eta_i,$$

where η_i is the number of Zs before and including the ith $+1$. η_i is called the rank of the ith order statistic in the Xs.

$D_{m,n}^+$ and $D_{m,n}$ (K–S statistic):

$$D_{m,n}^+ = \max_i \left(\frac{a_i}{m} - \frac{b_i}{n} \right),$$

$$D_{m,n} = \max_i \left| \frac{a_i}{m} - \frac{b_i}{n} \right|,$$

where

$$a_i = |\{j : j \le i, Z_j = 1\}|,$$
$$b_i = |\{j : j \le i, Z_j = -1\}|.$$

$G_{n,n}$ (Galton statistic):

$$2G_{n,n} = |\{j : \text{either } S_j > 0 \quad \text{or} \quad S_j = 0 \quad \text{and} \quad S_{j-1} = 1\}|,$$

where $S_j = Z_1 + \cdots + Z_j$. $2G_{n,n}$ represents the number of positive steps in the restricted random walk $R(n, n)$.

Let F_1 and F_2 be the continuous distribution functions of the two sets of random variables, respectively. The statistical problem in question is to ascertain whether or not two samples are from the same population (i.e., $F_1 = F_2$), and thus it is important to derive probability distributions of various statistics when $F_1 = F_2$ is true. Under the hypothesis $F_1 = F_2$ it can be shown ([24, p. 130; 61, p. 442]) that

$$P(Z_1 = z_1, \ldots, Z_{m+n} = z_{m+n}) = \frac{1}{\binom{m + n}{n}}, \qquad (4.24)$$

which is invariant for any $(z_1, \ldots, z_{m+n})$. This remarkable fact simply means that if we use the usual path representation of a sequence of $+1$s and -1s, each path from $(0, 0)$ to (m, n) is equally likely. Therefore the determination of the distribution of any rank order statistic under the hypothesis $F_1 = F_2$ (henceforth we assume this hypothesis) involves counting paths under a certain constraint. As an illustration the distribution of $U_{m,n}$ needs the number of paths to (m, n) having a given area underneath. Unfortunately, there is no single expression for this number; therefore in such situations one usually resorts to recurrence relations and prepares tables of distributions for different values of m, n up to a certain maximum, which it is possible to do with the aid of high speed computers. Because of (4.24) we may sometimes associate the statistic with the corresponding path or paths without any ambiguity. The path technique of obtaining the distribution may be called Gnedenko's technique, which was suggested in a paper of Gnedenko and Korolyuk [17] (also see [8]).

Various textbooks have treated the distribution of the aforementioned statistics, and thus we discuss only a few cases of interest that have been dealt with recently. Let us take the K–S statistics first. The events $(mnD_{m,n}^{+} \leq c)$ and $(mnD_{m,n} \leq c)$ correspond to the paths that do not cross the line $nx - my = c$ and the lines $nx - my = \pm c$, respectively. Now we readily obtain

$$\binom{2n}{n}P(nD_{n,n}^{+} \leq c) = \binom{2n}{n} - \binom{2n}{n+c+1}, \quad c \geq 0, \quad (4.25)$$

$$\binom{2n}{n}P(nD_{n,n} \leq c) = \sum_{j}\left[\binom{2n}{n - 2j(c+1)}_{+}\right.$$

$$\left. - \binom{2n}{n - (2j+1)(c+1)}_{+}\right] \quad (4.26)$$

from (1.3) and (1.7). Under the assumption that $F_1 (=F_2)$ is continuous except for a finite number of jumps, the distributions of $D_{n,n}^{+}$ and $D_{n,n}$ have been derived by Vincze [60], with the aid of (4.25) and (4.26). However, the result of Chapter 2 (Theorem 1) leads to

$$\binom{m+n}{n}P(mnD_{m,n}^{+} \leq c) = |(0, \mathbf{a}; 1)| \quad (4.27)$$

and

$$\binom{m+n}{n}P(mnD_{m,n} \leq c) = |(\mathbf{b}, \mathbf{a}; 1)| \quad (4.28)$$

(see notations of Chapter 2), where

$$b_i = \max\left(0, m - \left[\frac{c + (n-i)m}{n}\right]\right)$$

and

$$a_i = \min\left(m, \left[\frac{c + (i-1)m}{n}\right]\right), \quad i = 1, 2, \ldots, n.$$

Quite clearly we have derived the distributions and joint distributions of various statistics in the previous section if the random walk problem is reformulated in terms of statistics. Vellore [59] has successfully attempted the derivation of the joint distribution of K–S statistics and runs for $m = n$.

THEOREM 6

(a) $\binom{2n}{n} P(nD_{n,n}^+ \leq c, R_{n,n} = 2r)$

$$= \left[2\binom{n-1}{r-1}^2 - \binom{n+c-1}{r-1}\binom{n-c-1}{r-1} \right.$$

$$\left. - \binom{n+c-1}{r-2}\binom{n-c-1}{r} \right] \tag{4.29}$$

when $0 < c < n, 1 \leq r \leq n$,

(b) $\binom{2n}{n} P(nD_{n,n}^+ \leq c, R_{n,n} = 2r + 1)$

$$= 2\left[2\binom{n-1}{r-1}\binom{n-1}{r} - \binom{n+c-1}{r-1}\binom{n-c-1}{r} \right] \tag{4.30}$$

when $0 < c < n, 1 \leq r \leq n$.

Proof

In (a) we have only to show that the number of paths from $(0, 0)$ to (n, n) crossing the line $x = y + c$ and having $2r$ runs is given by

$$\binom{n+c-1}{r-1}\binom{n-c-1}{r-1} + \binom{n+c-1}{r-2}\binom{n-c-1}{r}. \tag{4.31}$$

Consider a path of such a type that starts with a horizontal step. Let $(s + c, s)$ be the last point on $x = y + c$ after which the path never touches the line. Construct a new path by first taking the reverse of the path from $(0, 0)$ to $(s + c, s)$ and then joining the reflection of the path from $(s + c, s)$ to (n, n) about the line $x = y + c$. The new path is seen to be a path from $(0, 0)$ to $(n + c, n - c)$ that starts with a vertical step and has $2r$ runs. The mapping is bijective and the number of latter paths can be derived as

$$\binom{n+c-1}{r-1}\binom{n-c-1}{r-1},$$

where the first (second) factor represents the number of ways in which the horizontal (vertical) units can be split into r groups.

Next take a path starting with a vertical step. Besides the point $(s + c, s)$, which was characterized earlier, let $(0, t)$ be the point at which the first horizontal step begins. By reversing the segment between $(0, t)$ and $(s + c, s)$ and reflecting the last part of the path about the line $x = y + c$, we get a path from $(0, 0)$ to $(n + c, n - c)$ with the following properties:

(i)   All horizontal units are divided into $r - 1$ groups and all vertical units into $r + 1$ groups;
(ii) the path starts with two groups of vertical units;
(iii) the remaining groups follow alternately, starting with a group with horizontal units.

The number of paths satisfying (i)–(iii) is

$$\binom{n + c - 1}{r - 2}\binom{n - c - 1}{r}.$$

Because of bijectivity we have completed the proof of (a). The proof of (b) is similar.

Vellore [59] also obtained the joint distribution of $D_{n,n}$ and $R_{n,n}$ by the same procedure. In another paper [54],† she derived the joint distribution of $R_{n,n}$ and another statistic $N^*_{n,n}$ (which may be called a crossing statistic), where $N^*_{n,n}$ is defined as

$$\left| \left\{ j: \sum_{i=1}^{j} Z_i = 0, \sum_{i=1}^{j-1} Z_i \sum_{i=1}^{j+1} Z_i = -1 \right\} \right|.$$

Denoting by $L^+_{m,n}$ the subscript j such that

$$na_j - mb_j > na_i - mb_i \qquad \text{for} \quad i < j$$

and

$$na_j - mb_j \geq na_i - mb_i \qquad \text{for} \quad i > j$$

(i.e., j is the subscript at which $D^+_{m,n}$ is first attained), Steck and Simmons [57] obtained the distribution of $L^+_{m,n}$ and the joint distribution of $(D^+_{m,n}, L^+_{m,n})$.

THEOREM 7

For $r = 1, \ldots, m + n$

$$\binom{m + n}{n} P(L^+_{m,n} = r) = |(a_1, \ldots, a_{m-1})|, \qquad (4.32)$$

† Miss Vellore's surname changed to Srivastava after her marriage.

where

$$a_i = \begin{cases} n - \left\{\dfrac{(m-i)n}{m}\right\} & \text{for} \quad i = 1, \ldots, \left\langle \dfrac{mr}{m+n} \right\rangle - 1, \\ n - \left\langle \dfrac{(m-i)n}{m} \right\rangle & \text{for} \quad i = \left\langle \dfrac{mr}{m+n} \right\rangle, \ldots, m-1, \end{cases}$$

and $\langle x \rangle$ and $\{x\}$ are, respectively, the smallest integer $\geq x$ and the smallest integer $> x$.

Proof

The theorem follows if we show that $N(L_{m,n}^+ = r)$ is equal to the number of paths from $(0, 0)$ to (n, m) that is never above the line $nx = my$ and moreover never touches the line after the $(m + n - r)$th step. This is proved by the following mapping. Consider the conjugate of any typical path corresponding to the event $(L_{m,n}^+ = r)$, which happens to be a path from $(0, 0)$ to (n, m) that touches the line $ny - mx = mnk$ for some k, $0 \leq k \leq 1$, for the first time at (s, t) on the line so that $s + t = r$. From this conjugate path construct a new path by first taking the segment from (s, t) to (n, m) and then joining to it the segment from $(0, 0)$ to (s, t). Evidently, this new path belongs to the set of paths stated at the beginning of this proof, and the mapping can be seen to be bijective when all possible values of ks are taken into account. This completes the proof.

COROLLARY

If m and n are relatively prime,

$$P(L_{m,n}^+ = r) = \frac{1}{m+n}. \tag{4.33}$$

Proof

Under the given restriction,

$$\left\langle \frac{nk}{m} \right\rangle = \left\{ \frac{nk}{m} \right\} \qquad \text{for} \quad k = 1, \ldots, m = 1.$$

Therefore $P(L_{m,n}^+ = r)$ is independent of r and then the corollary follows immediately.

It is easily noted from the correspondence in the proof that

$$N(L_{m,n}^+ = r) = N_{m,n} - \binom{m+n}{n} P(mnD_{m,n}^+ = 0, L_{m,n}^+ < r), \quad (4.34)$$

where $N_{m,n}$ represents the number of paths to (n, m) that is never above the line $nx = my$. This leads to the following theorem.

THEOREM 8

Let $u = \gcd(m, n)$ and $v = (m + n)/u$. Then

$$P(L_{m,n}^+ = r) \geq P(L_{m,n}^+ = r + 1) \qquad \text{for} \quad r = 1, \ldots, m + n - 1 \quad (4.35)$$

and

$$P(L_{m,n}^+ = hv + 1) = P(L_{m,n}^+ = hv + 2) = \cdots = P(L_{m,n}^+ = hv + v)$$
$$\text{for} \quad h = 0, 1, \ldots, u - 1. \qquad (4.36)$$

The next theorem gives the joint distribution of $D_{m,n}^+$ and $L_{m,n}^+$.

THEOREM 9

Let

$$b_i = n - x - \left\langle \frac{(m - y - i + 1)n}{m} \right\rangle, \qquad i = 1, \ldots, m - y,$$

and

$$d_i = \min\left(x, \left\langle \frac{ni}{m} \right\rangle - 1\right), \qquad i = 1, \ldots, y - 1.$$

Then

$$\binom{m+n}{n} P(mnD_{m,n}^+ = d, L_{m,n}^+ = r)$$
$$= \begin{cases} |(b_1, \ldots, b_{m-y})| \cdot |(d_1, \ldots, d_{y-1})|, \\ 0, \end{cases} \qquad (4.37)$$

according to whether or not there exists an integer solution to the equation $ny - mx = d, x + y = r$ such that $0 \leq x \leq n, 0 \leq y \leq m$.

The proofs of these two theorems are left as exercises. Here we may be reminded of the duality principle introduced in Chapter 1. In the proof of Theorem 7 if instead of taking the conjugate paths, we consider the paths themselves and use the construction, it would lead us to formulate a different expression.

From Chapter 3, Section 3 (the Chung–Feller theorem), it is well known that the distribution of $2G_{n,n}$ is uniform. The derivation of the joint distribution of the Galton statistic, with some others, has been initiated by Csáki and Vincze [6, 7], and we have covered some derivations in the last section.

In this section, as well as in the last, what is involved in determining the distributions amounts to counting lattice paths with restrictions. Thus we are led to employ several techniques to achieve our goal.

4. Rank Order Statistics—The Dwass Technique

Gnendenko's technique of finding the distributions of rank order statistics, as developed in the previous section, is purely of combinatorial nature and involves a random walk consisting of nonindependent steps in the sense that the number of $+1$s and -1s is given in advance. In contrast, the consideration of a random walk with independent steps forms the basis of the Dwass technique [12]. Moreover, the argument supporting the technique and the steps to be performed in it are essentially probabilistic, and thus it may be viewed as a probabilistic method applied to derive combinatorial results, more specifically those on lattice paths. Here one is reminded of our comments at the end of Chapter 2. A brief survey of the two methods is given by Šidak [52].

In order to describe the technique we limit ourselves to the case $m = n$ and start with the simple random walk in which $P(X_i = 1) = p$ and $P(X_i = -1) = 1 - p = q$ (say), $i = 1, 2, \ldots$. It is known ([16, p. 287]) that if $p < \frac{1}{2}$, the simple random walk is transient, and therefore with probability one there are only finitely many returns to the origin. Let

$$T = \max_j \left\{ j : S_j = 0 \quad \text{where} \quad S_j = \sum_{i=1}^{j} X_i \right\},$$

and let V be a function defined on the random walk. We say that V satisfies "assumption A" if V is completely determined by $S_1, \ldots, S_T$, $T > 0$. It is easy to establish that to each rank order statistic $V_{n,n}$ there corresponds a V satisfying "assumption A" such that the conditional distribution of V, given $T = 2n$, is exactly the distribution of $V_{n,n}$, and conversely. Here V is defined as

$$V(X_1, \ldots, X_T) = V_{n,n}(X_1, \ldots, X_{2n}),$$

where $T = 2n$ (the Ts can only be positive even integers). Because of the correspondence we use the dual notation V and $V_{n,n}$, where V is defined on the random walk satisfying "assumption A" and $V_{n,n}$ is the corresponding rank order statistic.

Let $h(p) = E(V)$, the expectation of V. For $p < \frac{1}{2}$

$$h(p) = E(V) = \sum_{n=0}^{\infty} E(V \mid T = 2n)P(T = 2n)$$

$$= \sum_{n=0}^{\infty} E(V_{n,n})P(T = 2n) .$$

In [16, p. 257] we have the generating function $\Psi(s)$ for the return time to the origin given as

$$\Psi(s) = 1 - (1 - 4pqs^2)^{1/2},$$

from which it is deduced that the probability of never returning to the origin is $1 - \Psi(1) = 1 - 2p$. Thus

$$P(T = 2n) = \binom{2n}{n}(pq)^n(1 - 2p), \tag{4.38}$$

and this leads to

$$\frac{h(p)}{1 - 2p} = \sum_{n=0}^{\infty} E(V_{n,n})\binom{2n}{n}(pq)^n, \tag{4.39}$$

which forms the main tool of the Dwass technique. Since $E(\phi(V_{n,n})) = P(V_{n,n} \text{ in } B)$, ϕ being the set indicator function of B, we take $\phi(V_{n,n})$ and $\phi(V)$ to find the distribution of the rank order statistic $V_{n,n}$. It is significant to remark that the usefulness of (4.39) depends on the ease with which one can explicitly evaluate $h(p)$ and then obtain the power series expansion $h(p)/(1 - 2p) = \sum_{n=0}^{\infty} a_n(pq)^n$, if available, so that we easily identify

$$\binom{2n}{n}E(V_{n,n}) = a_n. \tag{4.40}$$

We give two examples of the application of the Dwass technique in the following theorems.

THEOREM 10

Let

$$W_{n,n} = \left| \left\{ j: \sum_{i=1}^{j} Z_i = 0, j = 0, 1, \ldots, 2n \right\} \right|.$$

Then

$$\binom{2n}{n} p(W_{n,n} > k) = 2^k \binom{2n-k}{n-k}, \qquad n = k, k+1, \ldots. \quad (4.41)$$

Proof

The related function W on the restricted random walk represents the number of returns to the origin. More precisely,

$$W = |\{ j: S_j = 0, j = 0, 1, \ldots \}|.$$

Since the probability of ever returning to the origin is $\Psi(1) = 2p$, the probability of being at the origin more than k times is $(2p)^k$, which is identified as $h(p)$. But

$$p^k = P(X_1 = X_2 = \cdots = X_k = 1)$$

$$= \sum_{n=k}^{\infty} P(X_1 = \cdots = X_k = 1 \mid T = 2n) \binom{2n}{n} (pq)^n (1 - 2p) \quad \text{(by (4.38))}$$

$$= \sum_{n=k}^{\infty} \binom{2n-k}{n-k} (pq)^n (1 - 2p). \quad (4.42)$$

Thus

$$\frac{h(p)}{1-2p} = \frac{(2p)^k}{1-2p} = 2^k \sum_{n=k}^{\infty} \binom{2n-k}{n-k} (pq)^n.$$

Hence the theorem follows from (4.39).

THEOREM 11

$$\binom{2n}{n} P(nD_{n,n}^+ = k, L_{n,n}^+ = r) = \frac{k(k+1)}{r(2n-r+1)} \binom{r}{\frac{r+k}{2}} \binom{2n-r+1}{n - \frac{r+k}{2}}$$

$$(4.43)$$

if $r + k$ is even.

Proof

Let D^* be the supremum of the unrestricted walk. Clearly D^* corresponds to $nD_{n,n}^+$. Consider

$$\sum_{r=k}^{\infty} P(D^* = k, L^+ = r)s^r. \qquad (4.44)$$

The event $(D^* = k, L^+ = r)$ is equivalent to the compound event that

(i) the random walk reaches k for the first time in r steps and
(ii) it never reaches $k + 1$ afterward.

Moreover the generating function for the first passage time through k is $(\Psi(s)/2qs)^k$ and the probability of ever reaching one is $(\Psi(1)/2q) = (p/q)$ (see [16, p. 255]). Therefore

$$\sum_{r=k}^{\infty} P(D^* = k, L^+ = r)s^r = \frac{(1 - (1 - 4pqs^2)^{1/2})^k}{(2qs)^k}\left(1 - \frac{p}{q}\right). \qquad (4.45)$$

Representing the left-hand side of (4.45) by $h(p)$, we have

$$\frac{h(p)}{1 - 2p} = \frac{(1 - (1 - 4pqs^2)^{1/2})^k p^{k+1}}{2^k s^k (pq)^{k+1}}. \qquad (4.46)$$

In (4.42), we have an expansion of $p^k/(1 - 2p)$. Using this, an expansion of p^{k+1} is obtained. Quite clearly

$$p = \tfrac{1}{2}(1 - (1 - 4pq)^{1/2}).$$

The mapping $pq = t$ is $1 : 1$ from $[0, \tfrac{1}{2})$ onto $[0, \tfrac{1}{4})$, which changes (4.42) to

$$\left(\frac{1}{2}\right)^k \frac{(1 - (1 - 4t)^{1/2})^k}{(1 - 4t)^{1/2}} = \sum_{n=k}^{\infty} \binom{2n - k}{n - k} t^n. \qquad (4.47)$$

Integrating both sides with respect to t and determining the constant of integration from $t = 0$, we get

$$\left(\frac{1}{2}\right)^{k+1} (1 - (1 - 4t)^{1/2})^k = (k + 1) \sum_{n=k+1}^{\infty} \frac{1}{2n - k - 1}\binom{2n - k - 1}{n - k - 1} t^n. \qquad (4.48)$$

When t is replaced back by pq, (4.48) becomes

$$p^{k+1} = (k + 1) \sum_{n=k+1}^{\infty} \frac{1}{2n - k - 1}\binom{2n - k - 1}{n - k - 1}(pq)^n. \qquad (4.49)$$

Also, putting $t = pqs^2$ and substituting k for $k + 1$ give

$$\frac{(1 - (1 - 4pqs^2)^{1/2})^k}{2^k} = k \sum_{n=k}^{\infty} \frac{1}{2n - k} \binom{2n - k}{n - k} (pqs^2)^n. \quad (4.50)$$

Thus (4.46) takes the form

$$\frac{h(p)}{1 - 2p} = \left(k \sum_{i=k}^{\infty} \frac{1}{2i - k} \binom{2i - k}{i - k} (pqs^2)^i \right)$$

$$\times \left((k + 1) \sum_{j=k+1}^{\infty} \frac{1}{2j - k - 1} \binom{2j - k - 1}{j - k - 1} (pq)^j \right), \quad (4.51)$$

in which $\binom{2n}{n} P(nD_{n,n}^+ = k, L_{n,n}^+ = r)$ is the coefficient of $t^r(pq)^n$. This coefficient is easily checked to be the expression in the theorem, and hence the proof is complete.

By using the Dwass technique with more generating functions and proceeding as above, Aneja and Kanwar Sen [3, 4] have derived (4.18) and many other joint distributions.

The richness of the Dwass technique as a unified method cannot be realized fully because of the limitation on its use as pointed out earlier. For example, besides determining $h(p)$ explicitly, one has to obtain an explicit expression for the probability of never returning to the origin, which happens to be $1 - 2p$ when $m = n$ and $p < \frac{1}{2}$. This fact indicates that the general case of unequal sample sizes may not be easy to deal with when using the Dwass technique. However the technique is applicable in finding quite a few distributions for the case in which $m = \mu n$ [41], which case is possible because of the availability of the following two important relations [18] to be discussed in Chapter 6:

$$\sum_{k=0}^{\infty} \frac{\alpha}{\alpha + (\mu + 1)k} \binom{\alpha + (\mu + 1)k}{k} \theta^k = x^\alpha, \quad (4.52)$$

$$\sum_{k=0}^{\infty} \binom{\alpha + (\mu + 1)k}{k} \theta^k = \frac{x^{\alpha+1}}{\mu + 1 - \mu x}, \quad (4.53)$$

where

$$\theta = \frac{x - 1}{x^{\mu+1}} \quad \text{and} \quad |\theta| < \frac{\mu^\mu}{(\mu + 1)^{\mu+1}}.$$

The last inequality assures the convergence of the series. Relation (4.52) can be derived from Lagrange series expansion [see (1.19) and (1.21)] and relation (4.53) follows from (4.52) when we differentiate both sides with respect to θ.

The change introduced in the random walk by $m = \mu n$ is that X_i takes the value $+1$ or $-\mu$ with probability p or $1 - p = q$, respectively. Let $U(s)$ and $F(s)$ represent the generating functions for the return time to zero and the time for the first return to zero, respectively. Clearly

$$U(s) = \sum_{n=0}^{\infty} \binom{(\mu + 1)n}{n} p^{\mu n} q^n s^{(\mu+1)n} = \frac{z}{\mu + 1 - \mu z}$$

$$\text{(by (4.53)),} \qquad (4.54)$$

where

$$s^{\mu+1} p^{\mu} q = \frac{z - 1}{z^{\mu+1}} \qquad \text{and} \qquad |s|^{\mu+1} p^{\mu} q < \frac{\mu^{\mu}}{(\mu + 1)^{\mu+1}}$$

and

$$F(s) = \frac{U(s) - 1}{U(s)} = (\mu + 1) p^{\mu} q z^{\mu} s^{\mu+1} \qquad \text{([16, p. 285]).} \quad (4.55)$$

Therefore the probability of never returning to zero is given by

$$1 - F(1) = (\mu + 1) p^{\mu} q y^{\mu} \qquad (4.56)$$

with

$$p^{\mu} q y^{\mu+1} - y + 1 = 0 \qquad \text{and} \qquad p^{\mu} q < \frac{\mu^{\mu}}{(\mu + 1)^{\mu+1}}$$

(y is the value of z when $s = 1$). Note that if

$$p < \frac{\mu}{\mu + 1}, \qquad \text{then} \qquad p^{\mu} q < \frac{\mu^{\mu}}{(\mu + 1)^{\mu+1}},$$

in which case $F(1) < 1$; i.e., the event of returning to the origin is a transient recurrent event. Thus when $m = \mu n$ and $p < \mu/(\mu + 1)$, relation (4.39) is modified to

$$\frac{h(p)}{1 - (\mu + 1) p^{\mu} q y^{\mu}} = \sum_{n=0}^{\infty} E(V_{\mu n, n}) \binom{(\mu + 1)n}{n} p^{\mu n} q^n. \qquad (4.57)$$

The generalized expansion of p^k similar to (4.42) is

$$\frac{p^k}{1 - (\mu + 1) p^{\mu} q y^{\mu}} = \sum_{n=\langle k/\mu \rangle}^{\infty} \binom{(\mu + 1)n - k}{n - k} p^{\mu n} q^n, \qquad k > 0. \quad (4.58)$$

It may also be seen by using (4.52) that the probability generating function for the first passage time through one is psz, from which it follows that the probability generating function for the first passage through k is $(psz)^k$ and the probability of ever reaching one is py.

Proceeding as in the proof of Theorem 11 and using the above results for $m = \mu n$, we have the following generalization of Theorem 11.

THEOREM 12

$$\binom{(\mu + 1)n}{n} P(\mu n D^+_{\mu n, n} = k, L^+_{\mu n, n} = k + (\mu + 1)i)$$

$$= \frac{k}{k + (\mu + 1)i} \binom{k + (\mu + 1)i}{i} \left[\binom{(\mu + 1)(n - i) - k}{n - i} \right.$$

$$- \sum_{j=[k/\mu]+1}^{n-i} \binom{(\mu + 1)j - k - 1}{j} \frac{1}{1 + (\mu + 1)(n - i - j)}$$

$$\left. \times \binom{1 + (\mu + 1)(n - i - j)}{n - i - j} \right]$$

$$\tag{4.59}$$

for $i = 0, 1, \ldots, n - \langle k/\mu \rangle - 1, k = 0, 1, \ldots, n - 1$. The negative term is zero when $k = \mu n$ and $i = 0$.

It would be of interest to derive (4.59) from Theorem 9 as a special case.

Reimann and Vincze [47] have proposed test statistics that are functions of $(a_i - b_i)$ (for definitions see Section 3). If m and n are close, statistics of this type can be useful. Note that when $m = n$, K–S statistics are of this type. For this type of statistic we consider the associated random walk to be the simple random walk and T to be the time for the last return to r (without loss of generality we assume $m \geq n, m - n = r$). By introducing this change we can see that the background preparation for the application of the Dwass technique is ready. In this case (4.39) takes the form

$$\frac{E(V)}{1 - 2p} = \sum_{n=0}^{\infty} E(V_{n, n}) \binom{2n + r}{n} p^{n+r} q^n \tag{4.60}$$

for $p < \frac{1}{2}$. For the derivation of the distributions of some of these statistics see [2, 44].

As a final remark we observe that in applying the Dwass technique, in spite of repeated use of probabilistic tools, certain combinatorial results of an elementary nature have been utilized. Moreover, the technique can be thought of as a variation of the generating function method.

5. Concluding Remarks

While we may say that identities (4.2) and (4.3) are of the Vandermonde-type, their continuous analogues are the following Abel-type identities (see Gould [19, 20]; also Chapter 6, Section 2):

$$\sum_{k=0}^{n} \frac{a}{a + \mu k} \frac{(a + \mu k)^k}{k!} \frac{b}{b + \mu(n - k)} \frac{(b + \mu(n - k))^{n-k}}{(n - k)!}$$

$$= \frac{a + b}{a + b + \mu n} \frac{(a + b + \mu n)^n}{n!}, \tag{4.61}$$

$$\sum_{k=0}^{n} \frac{(a + \mu k)^k}{k!} \frac{b}{b + \mu(n - k)} \frac{(b + \mu(n - k))^{n-k}}{(n - k)!} = \frac{(a + b + \mu n)^n}{n!}. \tag{4.62}$$

That

$$\frac{a}{a + \mu n} \frac{(a + \mu n)^n}{n!}$$

is the measure of the set

$$\{(x_1, \ldots, x_n): 0 \le x_1 \le \cdots \le x_n, x_i \le a + \mu i$$
$$\text{for all } i, a \ge 0 \ge \mu \ge 0\}$$

follows from Theorem 6 of Chapter 2 (with some simplification), which is a natural generalization of the path number

$$\frac{a}{a + \mu n} \binom{a + \mu n}{n}$$

of (1.11). Remember that

$$\frac{a}{a + \mu n} \binom{a + \mu n}{n}$$

is a special case of (2.2).

Since paths are related to rank order statistics, so are the sets of the above type related to order statistics in an obvious way. In any standard textbook ([61, p. 236]) one can see that the joint probability density function f of the order statistics $X_{(1)}, \ldots, X_{(n)}$, from uniform distribution over $(0, 1)$, is given by

$$f(x_{(1)}, \ldots, x_{(n)}) = n!, \qquad 0 < x_{(1)} < \cdots < x_{(n)} < 1.$$

Therefore it follows that for $0 \leq u_i \leq v_i$, $i = 1, \ldots, n$,

$$P(u_i \leq X_{(i)} \leq v_i, i = 1, \ldots, n) = n! \qquad (4.63)$$

(measure of the set $\{(x_1, \ldots, x_n): 0 \leq x_1 \leq \cdots \leq x_n, u_i \leq x_i \leq v_i$ for all $i\}$), which can be evaluated from (2.41) and was obtained by Steck [55]. It may be noted that Epanechnikov [15] anticipated the general form through a recurrence relation. This result has many applications. We give one in which the distribution of a statistic similar to $D_{m,n}$ is derived. The empirical distribution function $F_n(x)$ at x of a random sample $X_1, \ldots, X_n$ is given by

$$F_n(x) = \frac{|\{j: x_j \leq x\}|}{n}.$$

If the sample is from the uniform distribution over $(0, 1)$, then the statistic D_n (Kolmogorov–Smirnov one-sample statistic) is defined as

$$D_n = \sup_{0 < x < 1} |F_n(x) - x|.$$

It is easy to verify that for $c > 0$

$$P(D_n \leq c) = n! \left\| \frac{(v_i - u_j)^{j-i+1}}{(j - i + 1)!} \right\|_{n \times n}, \qquad (4.64)$$

where

$$u_i = \begin{cases} 0 & \text{for} \quad i \leq nc, \\ \dfrac{i - nc}{n} & \text{for} \quad i \geq nc, \end{cases}$$

and

$$v_i = \begin{cases} \dfrac{nc + i - 1}{n} & \text{for} \quad i < n - nc + 1, \\ 1 & \text{for} \quad i > n - nc + 1. \end{cases}$$

Statistic D_n is used to test whether or not the sample is from a uniform distribution over $(0, 1)$. For finding the distributions of some statistics similar to D_n, in testing problems, and for other applications one may refer to Steck [55]. In fact he has evaluated the probability that the empirical distribution function for any continuous distribution lies between two distribution functions. In [21] an explicit expression for the generating function of D_n is obtained, from which a closed form expression for $P(D_n < u/n), u = 1, 2, \ldots$, is derived. A detailed study has been done by Durbin [9] on the distribution theory of statistics based on $F_n(x)$ such as D_n. Also see Gupta and Panchapakesan [23].

The relation between order statistics from uniform distribution and a Poisson process is well known. For example, given that n events have occurred during $(0, t)$ from a Poisson process, their successive times are jointly distributed as the order statistics of size n from the uniform distribution over $(0, t)$. Thus in addition to tests of goodness of fit we have another direction of application, viz., on a Poisson process (see Cox and Lewis [5]). Moreover, because of the above relations it is no surprise to find a continuous analogue of the Dwass technique, given by Dwass himself [13]. Also, a case of interest is the application of testing the serial independence in time series analysis as described by Durbin [10, 11].

Expression (4.63) may be viewed as the joint distribution of several special linear combinations of order statistics, more particularly those which involve the evaluation of the volume of a polyhedron determined by $0 \le x_1 \le \cdots \le x_n$ and $u_i \le x_i \le v_i$ for all i. If we consider general linear combinations of order statistics $\sum_{j=1}^{n} a_{ij} X_{(j)}, i = 1, \ldots, k$, then the distribution needs an expression for the volume of the polyhedron given by $0 \le x_1 \le \cdots \le x_n$ and $\sum_{j=1}^{n} a_{ij} x_j \le v_i, i = 1, \ldots, k$, which is rather more difficult to determine. Using the characteristic function technique, Ali and Mead [1] were able to obtain an explicit expression for the same, which, remarkably enough, contains determinants.

In Section 3 we found that the consideration of lattice paths is helpful in determining the distributions of various rank order statistics under the hypothesis $F_1 = F_2$. It is pleasantly surprising that the paths have some role to play when certain alternative hypotheses are considered. In this context assume that the probability density functions f_1 and f_2, corresponding to F_1 and F_2, exist, and let the class of alternative hypotheses be those for which the likelihood ratio f_2/f_1 is monotone, say increasing. Let $\mathbf{R} = (R_1, \ldots, R_m)$ be the rank vector of the X_is where R_i is the rank of the ith order statistic in the Xs (see Section 3). We say that $\mathbf{r}$ dominates $\mathbf{r}^1$ when the path $(r_1 - 1, r_2 - 2, \ldots, r_m - m)$ dominates the path vector $(r_1^1 - 1, r_2^1 - 2, \ldots, r_m^1 - m)$. Savage [49] has shown that if $\mathbf{r}$ dominates $\mathbf{r}^1$, then $P(\mathbf{R} = \mathbf{r}^1) > P(\mathbf{R} = \mathbf{r})$ under the above-mentioned alternatives. (For further refinement on ordering see [43].) Clearly the rejection region of admissible nonrandomization tests (for definitions see [61]) containing the rank vector $\mathbf{r}$ must contain all rank vectors dominated by $\mathbf{r}$. Hence we see that the size of the rejection region cannot be less than $T_m(\mathbf{r})/\binom{m+n}{n}$ $[= L(\mathbf{r})$, say], where $T_m(\mathbf{r})$ is the number of rank vectors dominated by $\mathbf{r}$, which is equal to the number of paths dominated by the path $(r_1 - 1, r_2 - 2, \ldots, r_m - m)$. $L(\mathbf{r})$ is called the lower significance probability bound. If any admissible test has size less than $L(\mathbf{r})$, then $\mathbf{r}$ is not in the rejection region for that test. A similar argument would give us $U(\mathbf{r})$, the upper significance probability bound such that any admissible test of size greater than $U(\mathbf{r})$ will have $\mathbf{r}$ in its rejection region. For a full discussion one is referred to Switzer's paper [58]. A natural analogy exists in the one-sample case, and the significance probability bounds can be computed under the null hypothesis of symmetry about zero. The basic ordering theorems are again due to Savage [50].

A second example due to Steck [56] is the computation of probability under the alternatives $F_1 = F_2^k, k > 0, F_2$ being uniform under $(0, 1)$. By using (4.63) he proves that

$$P(R_i \leq b_i, i = 1, \ldots, m | F_1 = F_2^k)$$

$$= \frac{n!}{(n + km)!} \left\| \binom{j}{j - i + 1} \frac{\Gamma(\theta_i + kj)}{\Gamma(\theta_i + ki - k)} \right\|_{m \times m}, \qquad (4.65)$$

where $\{b_i\}$ is an increasing sequence of integers and $\theta_i = b_i - i + 1$ for all i. We may again observe that the expression involves a determinant. Steck's proof needs knowledge of properties of the beta distribution. He has further generalized (4.65).

Because of our remark following (1.28), it is easy to see that the set of rank vectors forms a distributive lattice. Curiously enough, Savage [51] has given a lattice-theoretic interpretation of the rank sum statistic, which is as follows:

We say that rank vector $\mathbf{r}$ covers $\mathbf{s}$ if $\mathbf{r}$ dominates $\mathbf{s}$ and there does not exist another rank vector $\mathbf{t}$ which dominates $\mathbf{s}$ and is dominated by $\mathbf{r}$. A sequence of rank vectors $(\mathbf{s} = \mathbf{t}_0, \mathbf{t}_1, \ldots, \mathbf{t}_d = \mathbf{r})$ such that $\mathbf{t}_{i+1}$ covers $\mathbf{t}_i, i = 0, 1, \ldots, d - 1$, is called a chain from $\mathbf{s}$ to $\mathbf{r}$, and d is called the length of the chain. Denote by $d(\mathbf{r})$ the length of the chain from the rank vector $(1, 2, \ldots, m)$ to $\mathbf{r}$. Then $d(\mathbf{R})$ is related to $U_{m,n}$ as

$$d(\mathbf{R}) = U_{m,n} - \frac{m(m + 1)}{2}. \qquad (4.66)$$

Another interesting observation (see [45]) is that the probability generating function of $d(\mathbf{R})$ under $F_1 = F_2$ can be expressed in terms of the Gaussian binominal coefficient† $\binom{n}{r}_q$, as

$$\binom{m+n}{n} \sum_{k=0}^{mn} P(d(R) = k \,|\, F_1 = F_2)s^k = \binom{m+n}{n}_s, \qquad (4.67)$$

where

$$\binom{n}{r}_q = \frac{(q^n - 1)(q^{n-1} - 1) \cdots (q^{n-r+1} - 1)}{(q^r - 1)(q^{r-1} - 1) \cdots (q - 1)}. \qquad (4.68)$$

Note that when $q \to 1$,

$$\binom{n}{r}_q = \binom{n}{r}.$$

(A generalization of (4.67) which gives an expression for the probability generating function of the joint distribution of the K–S statistic and $d(\mathbf{R})$ has been done by the author jointly with B.R. Handa.)

The generating functions (4.52) and (4.53) have more applications than are apparent in this chapter. For example, in ascertaining the probability that not more than a given fraction of a community is infected with a particular disease, given that the community size has (a) geometric or (b) logarithmic series distribution, Rao [46] makes use of (4.52) and (4.53).

Although we have been able to give exact expressions for various distributions, we may be reminded of our earlier comment in Section 3 on the rank sum statistic $U_{m,n}$ and in Chapter 1, Section 5, that sometimes it might be either necessary or even easier to compute the distributions by the simple use of (1.14) or similar recurrence relations and the necessary boundary conditions. In fact tables have been constructed in this manner for providing distributions of K–S statistics for small samples [37, 38].

We conclude by emphasizing our remark in the introduction that only illustrations of various techniques for deriving distributions and joint distributions are presented in this chapter. However for more derivations of distributions by similar clever manipulations of generating functions and path constructions the reader may see the long list

---

† Readers interested in these coefficients are advised to refer to G. E. Andrews, The Theory of Partitions, *Encyclopedia of Mathematics and Its Applications*, Vol. 2. Addison-Wesley, Reading, Massachusetts, 1976.

of papers [14, 27, 28, 29, 30–35], which is by no means complete. For our satisfaction a few distributions are selected to appear as exercises. It may be pointed out that Kanwar Sen [31] has listed a few path operations (including duality) and applied them in obtaining certain distributions.

Exercises

1. Write the two-boundary identity analogous to (4.4).
2. Prove Theorem 1 by using a bijective mapping.
3. Show that (4.7) and (4.8) follow from (4.5) and that (4.7) simplifies to (4.9) when $\mu = 1$.
4. Prove (4.11).
5. Prove the equivalence relation (4.17).
6. Find the joint generating function of all five variables involved in $R(n; u, v; g, h)$.
7. Let $B(n; u, v; g, h)$ be the same event as $R(n; u, v; g, h)$ except that the final position is unspecified. Find the distribution of $B(n; u, v; g, h)$ [28].
8. Let $A_{2n, 2m}^{2k}(u^-, -v^-)$ denote the random walk initially at the origin which arrives at $2m$ $(0 < 2m \le u)$ on the $2n$th step after crossing the origin $2k$ times in the presence of absorbing barriers at u and $-v$ (u and v positive integers). Similarly, let $A_{2n, 2m}^{2k}(u^+, -v^-)$ denote the same random walk except that there is only one absorbing barrier at $-v$ and that the walk reaches or crosses u before reaching $2m$. Denoting by $A_s, \cdot(\cdot, \cdot)$ the generating function $\sum_{n=1}^{\infty} N(A_{2n}, \cdot(\cdot, \cdot))s^n$, show that [32]

    (a) $A_{s, 2m}^{2k}(u^-, -v^-) = (1 + w)w^{2k+m}(w_1 w_2)^k \dfrac{1 - w^{u-2m}}{1 - w^{u+1}}$,

    (b) $A_{s, 2m}^{2k}(u^+, -v^-) = (1 + w)w^{2k+m}$

    $$\times \left(1 - \frac{w_1^k(1 - w^{u-2m})}{1 - w^{u+1}}\right),$$

    where

    $$w = \frac{1 - \sqrt{1 - 4s}}{1 + \sqrt{1 - 4s}}, \qquad w_1 = \frac{1 - w^{u-1}}{1 - w^{u+1}},$$

    and

    $$w_2 = \frac{1 - w^{v-1}}{1 - w^{v+1}}.$$

9. Prove Theorem 5. Hence find an expression for the probability that a path with t diagonal steps ends at (m, n).

10. Let the line segment between two lattice points without any lattice point in between be called a step. If a step is parallel to the line $x = ty$, where $t \geq 0$ is an integer, we say that the step is of type S_t. (We have avoided the earlier definition because the present one is simpler in our special case.) Let p_1 be the probability of a horizontal step and $p_{t-\mu}$ be the probability of a step of type S_t, $t = 0, 1, \ldots, \mu$. As soon as the path reaches the line $x = \mu y + k$ the next step is either of type S_t, $t = 0, 1, \ldots, \mu - 1$ (referred to as the reflection of type D_t) with probabilities $q_{t-\mu}$, $t = 0, 1, \ldots$, $\mu - 1$, or else it terminates there with probability q_0. Show that [25] the probability that a path has r_i steps of type S_i ($i = 0, 1, \ldots, \mu$), k_i reflections of type D_i ($i = 0, 1, \ldots, \mu - 1$), and is absorbed at the end is

$$
\left[\binom{k_0 + \cdots + k_{\mu-1}}{k_0, \ldots, k_{\mu-1}} \frac{k + \sum_{i=0}^{\mu-1} (\mu - i) k_i}{k - \sum_{i=0}^{\mu-1} k_i + \sum_{i=0}^{\mu} (\mu + 1 - i) r_i} \right.
$$
$$
\left. \times \binom{k - \sum_{i=0}^{\mu-1} k_i + \sum_{i=0}^{\mu} (\mu + 1 - i) r_i}{r_0 - k_0, \ldots, r_{\mu-1} - k_{\mu-1}, r_\mu} \right]
$$
$$
\times \left[\prod_{i=0}^{\mu-1} \left(\frac{q_{i-\mu}}{p_{i-\mu}} \right)^{k_i} \prod_{i=0}^{\mu} p_{i-\mu}^{r_i} p_1^{k + \Sigma_{i=0}^{\mu} (\mu - i) r_i} q_0 \right].
$$

11. Prove Theorem 6(b).
12. Find the joint distribution of $(D_{n,n}, R_{n,n})$.
13. Find the joint distribution of $(R_{n,n}, N_{n,n}^*)$.
14. Prove Theorem 8 and Theorem 9.
15. Letting

$$
N_{n,n}^+(r) = \left| \left\{ j : \sum_{i=1}^{j} Z_i = r + 1, \sum_{i=1}^{j-1} Z_i = r \right\} \right|
$$

($N_{n,n}^+(r)$ may be called up-crossings of r), derive its distribution by both the Gnedenko and the Dwass techniques.

16. Show by both the Gnedenko and the Dwass techniques that

$$
\binom{2n}{n} P(nD_{n,n}^- < r, nD_{n,n}^+ < k) = \sum_{j=-\infty}^{\infty} \left[\binom{2n}{n + j(r + k)}_+ \right.
$$
$$
\left. - \binom{2n}{n + r + j(r + k)}_+ \right],
$$

where

$$D_{m,n}^- = \max_i \left(\frac{b_i}{n} - \frac{a_i}{m}\right)$$

(see the definition of $D_{m,n}^+$ in Section 3).

17. Verify Theorem 4 by using the Dwass technique [3].
18. Let

$$Q_{m,n}^+ = \left|\left\{j: \frac{a_j}{m} - \frac{b_j}{n} = D_{m,n}^+\right\}\right|.$$

By the Dwass technique, show that [4]

$$\binom{2n}{n} P(nD_{n,n}^+ = k, L_{n,n}^+ = r, Q_{n,n}^+ = u)$$

$$= \frac{k}{r}\binom{r}{\frac{k+r}{2}} \frac{k+u-1}{2n-r-u+1}\binom{2n-r-u+1}{n+\frac{k}{2}-\frac{r}{2}}.$$

19. Prove Theorem 12.
20. Verify (4.64).
21. Compute $L(\mathbf{r})$ and $U(\mathbf{r})$ for the case in which we have observed 7 Xs and 10 Ys and the rank vector of the Xs is $(2, 5, 6, 9, 10, 11, 16)$.
22. Discuss the probability bounds for a one-sample case from a continuous distribution for which we test the hypothesis that the probability density function is symmetric about zero.

References

1. Ali, M. M., and Mead, E. R., On the distribution of several linear combinations of order statistics from the uniform distribution, *Bull. Inst. Statist. Res. Training, Univ. of Dacca,* **3** (1969), 22–41.
2. Aneja, K. G., Random walk and rank order statistics, Ph.D. thesis submitted to Delhi University (1975).
3. Aneja, K. G., and Kanwar Sen, Random walk and distribution of rank order statistics, *SIAM J. Appl. Math.* **23** (1972), 276–287.
4. Aneja, K. G., and Kanwar Sen, Maxima in random walk and related rank order statistics, *Studia Sci. Math. Hungar.* **7** (1972), 425–428.
5. Cox, D. R., and Lewis, P. A. W., *The Statistical Analysis of Series of Events.* Methuen, London, 1966.
6. Csáki, E., and Vincze, I., On some problems connected with the Galton-test, *Publ. Math. Inst. Hungar. Acad. Sci.* **6** (1961), 97–109.

7. Csáki, E., and Vincze, I., Two joint distribution laws in the theory of order statistics, *Mathematica (Cluj)*, **5** (1963), 27–37.

8. Drion, E. F., Some distribution-free tests for the difference between two empirical cumulative distribution functions, *Ann. Math. Statist.* **23** (1952), 563–574.

9. Durbin, J., Tests for serial correlation in regression analysis based on the periodogram of least-squares residuals, *Biometrika* **56** (1969), 1–15.

10. Durbin, J., Tests of serial independence based on the cumulated periodogram, *Bull. Inst. Internat. Statist.* **42** (1969), 1039–1048.

11. Durbin, J., *Distribution Theory for Tests Based on the Sample Distribution Function.* SIAM, Philadelphia, 1973.

12. Dwass, M., Simple random walk and rank order statistics, *Ann. Math. Statist.* **38** (1967), 1042–1053.

13. Dwass, M., Poisson process and distribution free statistics. *Adv. in Appl. Probab.* **6** (1974), 359–375.

14. Engelberg, O., On some problems concerning a restricted random walk, *J. Appl. Probab.* **2** (1965), 396–404.

15. Epanechnikov, V. A., The significance level and power of the two-sided Kolmogorov test in the case of small sample sizes, *Theory Probab. Appl.* **13** (1968), 686–690.

16. Feller, W., *An Introduction to Probability Theory and Its Applications*, Vol. 1, 2nd ed., Wiley, New York, 1957.

17. Gnedenko, B. V., and Korolyuk, V. S., On the maximum discrepancy between two empirical functions, *Dokl. Akad. Nauk SSSR.* **80** (1951), 525–528 [English transl.: *Selected Transl. in Math. Statist. and Probab.*, IMS and AMS 1 (1961), 13–16].

18. Gould, H. W., Some generalizations of Vandermonde's convolution, *Amer. Math. Monthly* **63** (1956), 84–91.

19. Gould, H. W., Final analysis of Vandermonde's convolution, *Amer. Math. Monthly* **64** (1957), 409–415.

20. Gould, H. W., Generalization of a theorem of Jensen concerning convolutions, *Duke Math. J.* **27** (1960), 71–76.

21. Govindarajulu, Z., Alter, R., and Gragg, L. E., The exact distribution of the one-sample Kolmogorov statistic, *Trabajos Estadist. Investigation Oper.* **26** (1975), 407–431.

22. Gupta, H. C., Random walk in the presence of a multiple-function barrier, *J. Math. Sci.* **1** (1966), 18–29.

23. Gupta, S. S., and Panchapakesan, S., On order statistics and some applications of combinatorial methods in statistics, in *A Survey of Combinatorial Theory* (J. N. Srivastava, et al., eds.), North-Holland Publ., Amsterdam, 1973. 217–250.

24. Hájek, J., and Šidák, Z., *Theory of Rank Tests.* Academic Press, New York, 1967.

25. Handa, B. R., On a generalization of one dimensional random walks with a partially reflecting barrier, *Canad. Math. Bull.* **14** (1971), 325–332.

26. Handa, B. R., and Mohanty, S. G., On some distributions concerning a restricted random walk, *Studia Sci. Math. Hungar.* **4** (1969), 99–108.

27. Jain, G. C., On cumulative sums in one-dimensional symmetric random walk, *J. Indian Statist. Assoc.* **4** (1966), 73–85.

28. Jain, G. C., Joint distribution of intersections, ($\pm$) waves and ($\pm$) steps—I, *Proc. Nat. Inst. Sci. India A*, **32** (1966), 460–471.

29. Jain, G. C., Joint distribution of intersections, ($\pm$) waves and ($\pm$) steps—II, *J. Indian Statist. Assoc.* **6**, (1968), 46–48.

30. Kanwar Sen, On some combinatorial relations concerning the symmetric random walk, *Publ. Math. Inst. Hungar. Acad. Sci.* **9** (1965), 335–357.

31. Kanwar Sen, Paths of an odd number of steps with final position unspecified, *J. Indian Statist. Assoc.* **7** (1969), 107–135.

32. Kanwar Sen, Distribution of crossings in restricted paths, *Acta Math. Acad. Sci. Hungar.* **22** (1971), 23–36.

33. Kanwar Sen, Paths crossing two and three lines, *Acta Math. Acad. Sci. Hungar.* **22** (1971), 37–49.

34. Kanwar Sen, On certain probability models in random walk, *Calcutta Statist. Assoc. Bull.* **21** (1972), 71–76.

35. Kanwar Sen, Restricted random walk with crossings, positive and total returns in three lines enumerated, *Calcutta Statist. Assoc. Bull.* **22** (1973), 121–127.

36. Lehner, G., One-dimensional random walk with partially reflecting barrier, *Ann. Math. Statist.* **34** (1963), 405–412.

37. Marliss, G. S., and Zayachkowski, W., The Smirnov two-sample statistic $D_{m,n}(n \leq m \leq 20)$. Dep. of Mathematics, Essex College, Windsor, Ontario, 1962.

38. Massey, J., Jr., The distribution of the maximum deviation between two sample cumulative step functions, *Ann. Math. Statist.* **22** (1951), 125–128.

39. Mohanty, S. G., Some convolutions with multinomial coefficients and related probability distributions, *SIAM Rev.* **8** (1966), 501–509.

40. Mohanty, S. G., On some generalizations of a restricted random walk, *Studia Sci. Math. Hungar.* **3** (1968), 225–241.

41. Mohanty, S. G., and Handa, B. R., Rank order statistics related to generalized random walk, *Studia Sci. Math. Hungar.* **5** (1970), 267–276.

42. Mohanty, S. G., and Vellore, S., On some distributions of a generalized restricted random walk, *Colloquia Mathematica Societatis János Bolyai* (European Meeting of Statisticians, 1972), (J. Gani *et al.*, eds, 547–555. North-Holland Publ., Amsterdam, 1972.

43. Narayana, T. V., Savage, I. R., and Saxena, K. M. L., Young chains and rank orders, *Canad. J. Statist.* **6** (1978), 41–47.

44. Očka, I., Simple random walk and rank order statistics, *Aplikace Matematiky* **22** (1977), 272–290.

45. Pólya, G., Gaussian binomial coefficients and the enumeration of inversions, *Proc. 2nd Chapel Hill Conf. Combinatorial Mathematics and Its Applications, Univ. of North Carolina at Chapel Hill, 1970*, pp. 381–384.

46. Rao, B. R., A note on generalized Dwass' expansions, *J. Indian Statist. Assoc.* **12** (1974), 59–66.

47. Reimann, J., and Vincze, I., On the comparison of two samples with slightly different sizes, *Publ. Math. Inst. Hungar. Acad. Sci.* **5** (1960), 293–309.

48. Rohatgi, V. K., A combinatorial identity, Problem 70–21, *SIAM Rev.* **14** (1972), 166–169.

49. Savage, I. R., Contributions to the theory of rank order statistics: the two-sample case, *Ann. Math. Statist.* **27** (1956), 590–615.

50. Savage, I. R., Contributions to the theory of rank order statistics: the one-sample case, *Ann. Math. Statist.* **30** (1959), 1018–1023.

51. Savage, I. R., Contributions to the theory of rank order statistics: applications of lattice theory, *Rev. Internat. Statist. Inst.* **32** (1964), 52–64.

52. Šidak, Z., Applications of random walks in non-parametric statistics, *Bull. Inst. Internat. Statist. Proc. 39th Session* **45** (Book 3) (1973), 34–43.

53. Spitzer, F., *Principles of Random Walk*, Van Nostrand, Princeton, New Jersey, 1964.

54. Srivastava, S., Joint distributions based on runs and on the number of intersections, *Studia Sci. Math. Hungar.* **8** (1973), 211–224.

55. Steck, G. P., Rectangular probabilities for uniform order statistics and the probability that the empirical distribution function lies between two distribution functions, *Ann. Math. Statist.* **42** (1971), 1–11.

56. Steck, G. P., A new formula for $P(R_i \leq b_i, 1 \leq i \leq m \,|\, m, n, F = G^k)$, *Ann. Probab.* **2** (1974), 147–154.

57. Steck, G. P., and Simmons, G. J., On the distributions of $R^+_{m,n}$ and (D^+_{mn}, R^+_{mn}), *Studia Sci. Math. Hungar.* **8** (1973), 79–89.

58. Switzer, P., Significance probability bounds for rank orderings, *Ann. Math. Statist.* **35** (1964), 891–894.

59. Vellore, S., Joint distributions of Kolmogorov–Smirnov statistics and runs, *Studia Sci. Math. Hungar.* **7** (1972), 155–165.

60. Vincze, I., On Kolmogorov–Smirnov type distribution theorems, In *Nonparametric Techniques in Statistical Inference* (M. L. Puri, ed.), Cambridge Univ. Press, Cambridge, 1970, pp. 385–403.

61. Wilks, S. S., *Mathematical Statistics*, Wiley, New York, 1962.

5 Discrete Distributions, Queues, Trees, and Search Codes

1. Introduction

This chapter is an extension of Chapter 4 and is kept separate to avoid lengthiness. The other motivation is to keep rank order statistics and the associated random walk (truncated) together in one place. We have already noted that these statistics, and therefore their distributions, are needed in nonparametric testing problems.

In this chapter we shall first be concerned with probability models related to random walks and, in particular, derive waiting time distributions in the simple random walk, a correlated random walk, and some queueing processes. The waiting time distributions provide discrete probability models (the simplest of which is the binomial distribution) for many practical situations. In addition the parameters of random walk models are estimated. Using earlier results, certain sets of trees and search codes are counted. At the end of the chapter two more applications of path enumeration are cited.

As in Chapter 4 we may caution that prior exposure to statistics and queueing theory is necessary for an understanding of the portion on estimation and queues.

2. Discrete Distributions and a Correlated Random Walk

Among the probability distributions related to lattice paths, the simplest and probably the oldest one is the binomial distribution. In terms of the simple random walk as described at the beginning of Section 2 of Chapter 4 let the number of $+1$s and -1s be α and β at any stage. For μ an integer and r a positive integer (these restrictions may be relaxed) the probability that $\alpha - \mu\beta > r$ occurs for the first time may give rise to a probability distribution. For example, $\mu = 1$ and $\mu = 0$, respectively, lead to binomial and negative binomial distributions, whence the inequality sign is obviously replaced by the equality sign. Alternatively, we may think of another random walk in which the particle at any stage moves either $+1$ or $-\mu$ units with probabilities p or $1 - p$ ($=q$), respectively, and the class of probability distributions is given by the probabilities of the first passage time through r, $r = 1$, $2, \dots$. Here when $\mu = -1$ (which corresponds to the binomial distribution), we interpret the walk as having two types of steps of the same length, viz., $+1$, one with probability p, and the other with probability q. In order to compute the probabilities we may use the lattice path correspondence of the random walk. It is of course necessary to check whether or not the probabilities of the first passage time form a probability distribution. For instance, it is known that ([14, p. 321]) when $\mu = 1$, such probabilities become a probability distribution provided $p \geq q$. However it is easy to see that we always get a probability distribution for negative μ, in which case the first passage is sure to occur.

Let $f(k; \mu, r, p)$ denote the probability of the first passage time through r with $k - \mu$s. Since $\mu = -1$ gives rise to the binomial distribution,

$$f(k; -1, r, p) = \binom{r}{k} p^{r-k} q^k, \qquad k = 0, 1, \dots, r.$$

When $\mu < -1$, the computation of $f(k; \mu, r, p)$ will involve numbers which form a partition of generalized Fibonacci numbers (see Section 4 of Chapter 1). It is also well known that the resulting distribution for $\mu = 0$ is negative binomial, so that

$$f(k; 0, r, p) = \binom{r + k - 1}{k} p^r q^k, \qquad k = 0, 1, \dots,$$

where $\binom{r+k-1}{k}$ is the number of paths from the origin to (r, k) that do not touch the line $x = r$ except at the end. Since the number of paths from

$(0, 0)$ to $(r + \mu k, k)$ that do not touch the line $x = \mu y + r$ except at the end is

$$\frac{r}{r + (\mu + 1)k} \binom{r + (\mu + 1)k}{k} \tag{5.1}$$

(Section 4, Chapter 1), we obtain

$$f(k; \mu, r, p) = \frac{r}{r + (\mu + 1)k} \binom{r + (\mu + 1)k}{k} p^{r + \mu k} q^k, \tag{5.2}$$

$k = 0, 1, \ldots$, which defines a probability distribution provided

$$\sum_{k=0}^{\infty} f(k; \mu, r, p) = 1. \tag{5.3}$$

In the power series expansion (4.52), when we put $x = 1/p$, we note that (5.3) is true provided

$$p^\mu q < \frac{\mu^\mu}{(\mu + 1)^{\mu + 1}}$$

[see the remark after (4.53)]. In fact, a further examination reveals that (5.2) defines a probability distribution even if μ and r are real numbers such that (5.2) is nonnegative for every k and (5.3) is satisfied. The distribution given by (5.2) may be regarded as either a natural generalization of the negative binomial distribution through the above random walk model (equivalently through the lattice paths) or one that can be derived directly from Lagrange's formula as was done by Jain and Consul [25]. Let the new distribution be called the generalized negative binomial (in short GNB) distribution with parameters (μ, r, p). Using the Lagrange formula approach, Consul and Shenton have generated a class of distributions called Lagrange distributions and studied their properties [8, 9, 52].

Again from (4.52), we can see that the probability generating function $G(s)$ of the GNB distribution satisfies

$$sq(G(s))^{(\mu + 1)/r} - (G(s))^{1/r} + p = 0. \tag{5.4}$$

Clearly, by repeated differentiation of (5.4) with respect to s at $s = 1$ we are able to obtain the factorial moments of the distribution from which the ordinary moments or the central moments may be obtained. The mean and variance are equal to

$$\frac{rq}{1 - (\mu + 1)q} \quad \text{and} \quad \frac{rpq}{(1 - (\mu + 1)q)^3}, \tag{5.5}$$

respectively. When r and μ are large while p is small and such that $rp = \lambda_1$ and $r\mu = \lambda_2$, λ_1 being finite and positive and $|\lambda_2| < 1$, the GNB distribution can be approximated as

$$\frac{\lambda_1(\lambda_1 + k\lambda_2)^{k-1}e^{-(\lambda_1 + k\lambda_2)}}{k!}, \tag{5.6}$$

which may be regarded as the generalized Poisson distribution (see [6, 7].)

From the convolution identity (4.2) we infer that the sum of two GNB random variables with parameters (μ, r_1, p) and (μ, r_2, p) is also a GNB random variable with parameters $(\mu, r_1 + r_2, p)$. The same identity also gives rise to a probability distribution analogous to the hypergeometric distribution given by

$$g(k; \mu, r_1, r_2, n) = \frac{\dfrac{r_1}{r_1 + (\mu + 1)k}\dbinom{r_1 + (\mu + 1)k}{k} \times \dfrac{r_2}{r_2 + (\mu + 1)(n - k)}\dbinom{r_2(\mu + 1)(n - k)}{n - k}}{\dfrac{r_1 + r_2}{r_1 + r_2 + (\mu + 1)n}\dbinom{r_1 + r_2 + (\mu + 1)n}{n}}. \tag{5.7}$$

For other properties and characterizations see [25]. The derivation of similar properties and generation of new distributions are possible on the basis of identities (4.3) and (4.4).

Suppose we want to obtain the minimum variance unbiased (MVU) estimator of p. Denoting by Z the sample space of outcomes of the first passage time in terms of lattice paths and defining the random variable K on Z to represent the number of vertical units, we see that K is a minimal sufficient statistic (Zacks [61, p. 49]), the distribution of which is seen to be complete (for a definition see [61, p. 68]). We look for an unbiased estimator of p. Let

$$Y = \begin{cases} 1 & \text{if the first trial is a horizontal unit,} \\ 0 & \text{otherwise.} \end{cases} \tag{5.8}$$

Then $E(Y) = p$ and Y is an unbiased estimator of p. By the well-known Lehmann–Scheffé theorem ([61, p. 104]) the unique MVU estimate of p

is given by $E(Y|k)$. Its evaluation depends on the number of paths from $(0, 0)$ to $(r + \mu k, k)$ that

  (i)   do not touch the line $x = \mu y + r$ except at the end,
  (ii) contain k vertical units, and
  (iii) start with a horizontal unit,

which is equal to

$$\frac{r - 1}{r - 1 + (\mu + 1)k} \binom{r - 1 + (\mu + 1)k}{k}, \quad r > 1.$$

Thus, the unique MVU estimate of p when $r > 1$, is

$$\frac{(r - 1)(r + \mu k)}{r(r - 1 + (\mu + 1)k)}. \tag{5.9}$$

It may be remarked that from the point of view of estimation each path constitutes a sequential sampling plan on a simple random walk. More on sequential sampling plans is given in Section 5.

Since there exists a multinomial generalization of (5.1) in terms of lattice paths, the multivariable generalization of (5.2) is seen to be [23, 32]

$$f(k_1, \ldots, k_j; \mu_1, \ldots, \mu_j, r; p_1, \ldots, p_j)$$

$$= \frac{r}{r + \sum (\mu_i + 1)k_i} \binom{r + \sum (\mu_i + 1)k_i}{k_1, \ldots, k_j} (1 - \sum p_i)^{r + \sum \mu_i k_i} \prod_{i=1}^{j} p_i^{k_i},$$

$$\tag{5.10}$$

where $\sum$ stands for $\sum_{i=1}^{k}$. Discussion on the generalized multivariate Poisson distribution and hypergeometric distribution and estimation is obvious.

In our approach negative binomial distribution (i.e., when $\mu = 0$) arises in a chance mechanism as a waiting time distribution. The interested reader is referred to Boswell and Patil [1] for other chance mechanisms in which the same distribution arises. For negative multinomial distribution see [26, 53]. Thus we are furnishing another situation of possible probabilistic treatments to a combinatorial problem. It is also recommended that the list of distributions given in [27] be consulted, which distributions are quite similar to those we have discussed above.

It is evident from the convolution identity (4.2) that the random variable for which (5.2) gives the probability distribution can be seen as the sum of r i.i.d. random variables (say, $X_1, \ldots, X_r$), each having a probability distribution the same as that in (5.2) with $r = 1$. Viewed this way, (5.9) is the unique MVU estimate of p, based on a random sample of size r from $f(k; \mu, 1, p)$, because it is known that $X_1 + \cdots + X_r$ is a complete sufficient statistic. Consider the decapitated (or truncated) probability distribution, given by

$$f^*(k; \mu, 1, p) = \alpha \frac{1}{(\mu + 1)k + 1} \binom{(\mu + 1)k + 1}{k} p^{\mu k + 1} q^k, \quad (5.11)$$

for $k = 1, 2, \ldots$, where

$$\alpha = \left(\sum_{k=1}^{\infty} \frac{1}{(\mu + 1)k + 1} \binom{(\mu + 1)k + 1}{k} p^{\mu k + 1} q^k \right)^{-1} = \frac{1}{1 - p}.$$

Clearly we can no longer use (4.2) to obtain the distribution of the sum of i.i.d. decapitated GNB variables, and as such the combinatorics related to decapitated distributions become different (see [2]). However in our particular case we can use Lagrange's formula to find the distribution of the sum, as has been done by Gupta [20].

Next we think of a correlated random walk for which the movement of the particle at any stage is not independent of the previous steps. Consider the situation in which the probabilities of moving in the positive direction at any stage are p_1 or p_2 according to whether or not the particle moved in the positive or negative direction in the previous stage. In other words the movements are governed by the transition probability matrix

		To	
From		Positive	Negative
Positive		p_1	q_1
Negative		p_2	q_2

such that $p_1 + q_1 = p_2 + q_2 = 1$. We assume $p_1 \neq p_2$; otherwise the case would be the same as that covered in our earlier discussion. This problem can be reformulated as a coin tossing game as was done in [44, 45]. As a particular case set $\mu = 1$. Let $f_1(r, n)$ $(f_2(r, n))$ denote the conditional probability of the first passage time through r in n steps, given that the particle initially arrived at the origin from -1 $(+1)$. Let

$G_i(r; s)$ be the probability generating function (in short, p.g.f.) of $f_i(r, n)$, $i = 1, 2$. Since $f_i(r, n) = 0$ when $n \neq 2k + r$, where k is the number of -1s in the sequence, we have

$$G_i(r; s) = \sum_{k=0}^{\infty} f_i(r, 2k + r)s^{2k+r}. \tag{5.12}$$

In order to obtain the distribution we may use the technique ([14, p. 318]) and observe that for $i = 1, 2$

$$f_i(r, n) = p_i f_1(r - 1, n - 1) + q_i f_2(r + 1, n - 1), \qquad n > r > 1, \tag{5.13}$$

with boundary conditions

$$\begin{aligned} f_i(1, 1) &= p_i, \\ f_i(1, n) &= p_i + q_i f_2(2, n - 1), \qquad n > 1, \\ f_i(r, r) &= p_i f_1(r - 1, n - 1). \end{aligned} \tag{5.14}$$

These lead to the recurrence relations

$$G_i(r; s) = p_i s G_1(r - 1; s) + q_i s G_2(r + 1; s), \qquad i = 1, 2, \tag{5.15}$$

with

$$G_1(0; s) = G_2(0; s) = 1. \tag{5.16}$$

Furthermore,

$$G_1(r; s) = (G_1(1; s))^r \tag{5.17}$$

and

$$G_2(r; s) = G_2(1; s)(G_1(1; s))^{r-1}. \tag{5.18}$$

Without any ambiguity we may write $G_i(1; s)$ briefly as $G_i(s)$. Using (5.15)–(5.18), we get

$$q_2 s(G_1(s))^2 + (q_1 p_2 s^2 - p_1 q_2 s^2 - 1)G_1(s) + p_1 s = 0 \tag{5.19}$$

and

$$q_1 s(G_2(s))^2 + (p_1 q_2 s^2 - q_1 p_2 s^2 - 1)G_2(s) + p_2 s = 0, \tag{5.20}$$

the solutions of which are

$$G_1(s) = \frac{1 + p_1 q_2 s^2 - q_1 p_2 s^2 - \sqrt{(1 + p_1 q_2 s^2 - q_1 p_2 s^2)^2 - 4p_1 q_2 s^2}}{2q_2 s} \tag{5.21}$$

and

$$G_2(s) = \frac{1 + p_2 q_1 s^2 - p_1 q_2 s^2 - \sqrt{(1 - p_2 q_1 s^2 - p_1 q_2 s^2)^2 - 4 p_2 q_1 s^2}}{2 q_1 s};$$

(5.22)

the other solution in each case being inadmissible so as to make $G_i(s)$ a p.g.f. Moreover $G_1(1) = G_2(1) = 1$ when $p_1 + p_2 > 1$, and thus under this condition the conditional probabilties $\{f_1(1, n)\}$ and $\{f_2(1, n)\}$ of first passage time through 1 (in fact also through r) give rise to probability distributions. Formally, these probabilities can be determined from the power series expansions of (5.21) and (5.22), and the actual computation could be tedious. However, as an alternative approach we could obtain the distributions explicitly by using lattice paths in a way that constitutes the minimal sufficient partition of the sample space of outcomes.

THEOREM 1

For $n = 2k + r$

$$f_1(r, n) = \begin{cases} p_1^r & \text{when} \quad k = 0, \\[2mm] \sum_{m=1}^{k} \left[\binom{k+r-1}{m} \binom{k-1}{m-1} - \binom{k+r-1}{m-1} \binom{k-1}{m} \right] \\ \qquad \times p_1^{k+r-m} p_2^m q_1^m q_2^{k-m} & \text{when} \quad r > 1, \quad k \neq 0, \\[2mm] \sum_{m=1}^{k} \left[\binom{k-1}{m-1} \binom{k-1}{m-1} - \binom{k-1}{m-2} \binom{k-1}{m} \right] \\ \qquad \times p_1^{k+1-m} p_2^m q_1^m q_2^{k-m} & \text{when} \quad r = 1, \quad k \neq 0, \end{cases}$$

(5.23)

and

$$f_2(r, n) = \begin{cases} \sum_{m=1}^{k+1} \left[\binom{k+r-2}{m-1} \binom{k}{m-1} - \binom{k+r-2}{m-2} \binom{k}{m} \right] \\ \qquad \times p_1^{k+r-m} p_2^m q_1^{m-1} q_2^{k-m+1} & \text{when} \quad r > 1, \\[2mm] \sum_{m=1}^{k+1} \left[\binom{k-1}{m-1} \binom{k-1}{m-1} - \binom{k-1}{m-2} \binom{k-1}{m} \right] \\ \qquad \times p_1^{k+1-m} p_2^m q_1^{m-1} q_2^{k-m+1} & \text{when} \quad r = 1. \end{cases}$$

(5.24)

Proof

The first part of (5.23) is direct. Let us first consider $f_1(r, 2k + r)$, where $k > 0$ and $r > 1$. Representing the random walk by lattice paths in the usual manner, we observe that a sequence in the event is represented by a path from $(0, 0)$ to $(2k + r, k)$, not touching the line $x = y + r$ except at the end. But each such path does not have the same probability. Let us call a sequence of a vertical unit followed by a horizontal unit a turn. It can be seen that each of those paths that consists of m turns, $m = 1, \ldots, k$ has the same probability, viz., $p_1^{k+r-m} \times p_2^m q_1^m q_2^{k-m}$. What remains to complete the proof is to count such paths. Taking the reverse paths into account, we immediately recognize the situation to be that of the variation of Theorem 3 in Chapter 2, where $n = 2$, $v_1 = k + t$, $t = 0, 1, \ldots, r - 1$, $v_2 = k$, and $k = m$. Therefore the number of paths is

$$\sum_{t=0}^{r-1} \left[\binom{k+t-1}{m-1} \binom{k-1}{m-1} - \binom{k+t-1}{m-2} \binom{k-1}{m} \right]$$
$$= \binom{k+r-1}{m} \binom{k-1}{m-1} - \binom{k+r-1}{m-1} \binom{k-1}{m},$$

and this checks with the second part of (5.23). When $r = 1$, the last part follows.

For $f_2(r, 2k + r)$, $r > 1$, we similarly note that any path

(i)   either starting with a vertical unit and having m turns or
(ii) starting with a horizontal unit and with $m - 1$ turns,

has the probability $p_1^{k+r-m} p_2^m q_1^{m-1} q_2^{k-m+1}$. When $r = 1$, (ii) does not arise. As before the number of paths in (i) is

$$\binom{k+r-2}{m-1} \binom{k-1}{m-1} - \binom{k+r-2}{m-2} \binom{k-1}{m}$$

and in (ii) is

$$\sum_{t=0}^{r-2} \left[\binom{k+t-1}{m-2} \binom{k-1}{m-2} - \binom{k+t-1}{m-3} \binom{k-1}{m-1} \right]$$
$$= \binom{k+r-2}{m-1} \binom{k-1}{m-2} - \binom{k+r-2}{m-2} \binom{k-1}{m-1}.$$

From this (5.24) follows, and the proof is complete.

The evaluation of mean and variance may perhaps be done directly from the distribution. But it seems simpler to get these from the recurrence relations of p.g.f. by differentiation as before. Unlike the uncorrelated random walk, where the moments are computed for the number of -1s, here we may find the moments for the duration of the walk. (Of course one can obtain either from the other.) Corresponding to $\{f_1(r, n)\}$, the mean and variance are

$$\frac{r(p_2 + q_1)}{p_2 - q_1} \quad \text{and} \quad \frac{4rp_2q_1(q_2 + p_1)}{(p_2 - q_1)^3}, \tag{5.25}$$

and corresponding to $\{f_2(r, n)\}$, these are

$$\frac{1}{p_2 - q_1}(p_1 + q_2 + (r - 1)(p_2 + q_1))$$

and $\tag{5.26}$

$$\frac{4}{(p_2 - q_1)^3}(q_2 p_1(p_2 + q_1) + (r - 1)p_2 q_1(q_2 + p_1)).$$

As in the one-parameter case, we would like to get MVU estimates of p_1 and p_2. Consider the probability distributions $\{f_1(r, n)\}$. Denoting by K and M the random variables representing the number of vertical units and turns, respectively, we may observe that (K, M) is a pair of minimal sufficient statistics and that its distribution is complete, which follow from [51].

Furthermore, if Y denotes the same random variable as in (5.8), but on the new sample space induced by $\{f_1(r, n)\}$, then Y is an unbiased estimator of p_1. Now for given k and m the number of paths which begin with a horizontal unit is

$$\binom{k + r - 2}{m}\binom{k - 1}{m - 1} - \binom{k + r - 2}{m - 1}\binom{k - 1}{m}, \qquad r > 1,$$

and thus for $r > 1$ the MVU estimate of p_1 is

$$E(Y \mid k, m) = \frac{\binom{k + r - 2}{m}\binom{k - 1}{m - 1} - \binom{k + r - 2}{m - 1}\binom{k - 1}{m}}{\binom{k + r - 1}{m}\binom{k - 1}{m - 1} - \binom{k + r - 1}{m - 1}\binom{k - 1}{m}}$$

$$= \frac{(r - 1)(k + r - m)}{r(k + r - 1)}. \tag{5.27}$$

The case $r = 1$ can be dealt with separately as before. When $k = 0$, $f_1(r, n) = p_1^r$, and therefore p_2 is not estimable.

For the distribution $\{f_2(r, n)\}$ we define the random variable M^*, representing the number of turns when the path starts with a vertical unit and the number of turns minus one when the path starts with a horizontal unit. Once again it can be verified that (K, M^*) is sufficient and complete. Moreover, Y is the unbiased estimator of p_2. Also, defining

$$Y^* = \begin{cases} 1 & \text{if the first horizontal unit is followed by another,} \\ 0 & \text{otherwise,} \end{cases}$$

we can easily check that $E(Y^*) = p_1$. Therefore we obtain the MVU estimators of p_2 and p_1 as

$$E(Y|k, m^*) = \frac{\binom{k + r - 2}{m - 1}\binom{k - 1}{m - 2} - \binom{k + r - 2}{m - 2}\binom{k - 1}{m - 1}}{\binom{k + r - 2}{m - 1}\binom{k}{m - 1} - \binom{k + r - 2}{m - 2}\binom{k}{m}}, \quad r > 1$$

(5.28)

and

$$E(Y^*|k, m^*) = \frac{\binom{k + r - 3}{m - 1}\binom{k}{m - 1} - \binom{k + r - 3}{m - 2}\binom{k}{m}}{\binom{k + r - 2}{m - 1}\binom{k}{m - 1} - \binom{k + r - 2}{m - 2}\binom{k}{m}}, \quad r > 2,$$

(5.29)

respectively. Note that without causing any ambiguity we have retained the same notation for random variables k and y, even though they are defined on different sample spaces.

Although our discussion has been limited to $\mu = 1$, similar results can be obtained for general $\mu \geq 0$ in an analogous way [33]. The case $\mu < 0$ has also been considered [15, 34, 45].

For better insight into the combinatorial structure of the probability distribution in Theorem 1 we present the coin-tossing game version of the random walk with $r = 2$. Consider two coins 1 and 2 such that the probability of obtaining a head with coin i is $p_i, i = 1, 2$. For equivalence with $f_2(r, n)$ the game should satisfy the following rules:

(a) Start with coin 1;
(b) the kth toss is made with coin 1 or 2 according as the $(k - 1)$st toss is a tail or a head;

(c)   stop tossing when for the first time the total number of heads
exceeds the total number of tails by exactly 2.

A sequence of outcomes is given below (H and T stand for a head and a
tail, respectively):

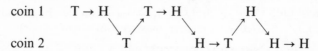

(The arrows may seem redundant.) In this sequence there are ten trials
of which five are made with coin 1.

We say that a sequence belongs to $S_n, n \geq 0$ if there are n Ts with
coin 1. Our example is a member of S_2. Notice that if we omit either

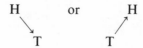

or both, the sequence still belongs to S_2, and this observation is indeed
true for any $n, n \leq 0$. Thus for convenience we call

a subsidiary sequence (briefly s.s.) and call the sequence after omission
of all s.s.s a base sequence (briefly b.s.). In the above case the b.s. is

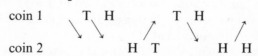

The only other b.s. of S_2 is

coin 1   T   H   T   H
coin 2   H   T   H   H

Observe that any number of s.s.s can be inserted at places indicated by
arrows in each b.s. of S_2 and that the number of such places is five. It can
be checked that in any b.s. of S_n one can insert s.s.s in $2n + 1$ places.

One way of distinguishing the above two b.s.s is that in one case
two Ts with coin 1 are together and in the other these are intercepted by
outcomes of coin 2 (i.e., Ts are partitioned into two parts). We say that a
b.s. of S_n has k partitions if Ts with coin 1 are partitioned into k parts,

$k = 1, \ldots, n$. It can be proved that the number of b.s.s of S_n having k partitions is the well-known number

$$\frac{1}{n} \binom{n}{k} \binom{n}{k-1}$$

(see Chapter 2, after (2.21)), so that the probability of sequences generated by b.s.s of S_n having k partitions is equal to

$$\frac{1}{n} \binom{n}{k} \binom{n}{k-1} \frac{q_1^n p_1^k p_2^{n+2} q_2^{k-1}}{(1 - p_1 q_2)^{2n+1}}, \qquad k = 1, \ldots, n, \ n = 0, 1, \ldots. \quad (5.30)$$

This result is due to Narayana [42]. For similar results on general r (i.e., $r \neq 2$) see [46].

Instead of conditional distributions, Jain [24] has assumed that the particle initially has reached the origin from -1 or $+1$ with probability ρ or $1 - \rho$ and has derived the unconditional distribution. A correlated random walk in the presence of two barriers, when $\mu = 1$, is dealt with in [19].

3. Queues

In his book Takács [58] has presented an excellent treatment of applications of combinatorial methods in the fluctuation of processes, in particular, queueing processes. Interested readers should find it useful to get acquainted with the materials in [58], as well as in Takács' other papers [54, 55, 56, 57]. Indeed what follows in this section is a direct development of his approach and, in a way, complements his work. However we give a few examples of certain aspects of queues involving batches, dealing mainly with the number of customers served in a busy period, which are directly related to lattice path combinatorics. In doing so we neither claim that other applications are not possible nor do we answer every question that may arise from our discussion.

We start with the queueing system $M/M/1$ [35] in which the customers arrive at a counter in batches of size m in accordance with a Poisson process of density λ_1 and are served individually by a single server. The service times are i.i.d. random variables with distribution function

$$F(x) = \begin{cases} 1 - e^{-\lambda_0 x} & \text{if} \quad x \geq 0, \\ 0 & \text{otherwise} \end{cases}$$

and are independent of the arrival times. The server is busy if and only if there is a single customer at the counter. Denote by $G(j, k, u, m; t)$ the probability that a busy period, initiated by $j \geq 1$ customers, has a maximum queue length $\leq k$, consists of u customers, and has length $\leq t$. Let $g(j, k\ u, m) = G(j, k, u, m; \infty)$.

THEOREM 2

$$g(j, k, u, m) = 0 \qquad \text{when} \quad k < j \quad \text{or} \quad u \neq mn + j;$$

$$n, \text{ a nonnegative integer.}$$

For $k \geq j$

$$g(j, k, mn + j, m) = |(\mathbf{b}, \mathbf{a}; 1)| p^n q^{nm+j}, \tag{5.31}$$

where

$$p = 1 - q = \frac{\lambda_1}{\lambda_0 + \lambda_1}$$

and $|(\mathbf{b}, \mathbf{a}; 1)|$ is as in Chapter 2 (Section 3) with

$b = (m + 1, 2m + 1, \ldots, mn + 1),$

$a = (k + 1, k + m + 1, \ldots, k + (i - 1)m + 1, mn + j, \ldots, mn + j).$

Proof

That the distribution of service times is exponential implies that the departure times form a Poisson process of density λ_0. Because the arrival and departure times are independent Poisson processes of density λ_1 and λ_0, respectively, one can consider that at the beginning of the busy period a Poisson process of density $\lambda_0 + \lambda_1$ starts and that every event in the process independent of others is either an arrival of a batch of size m with probability $p = \lambda_1/(\lambda_0 + \lambda_1)$ or a departure of a customer with probability $q = 1 - p$.

Suppose that at any instant the number of arrivals up to this time has been v and the number of departures has been w. Then the busy period continues as long as $w < mv + j$ and ends when for the first time $w = mv + j$. It consists of $mn + j$ services if and only if among the first

$n(m + 1) + j$ events in the Poisson process there would be n arrivals and $mn + j$ departures such that $w < mv + j$ except at the end.

Represent a departure by a horizontal unit and an arrival by a vertical unit. Then the busy period, consisting of $mn + j$ services (or departures), corresponds to the set of lattice paths from $(0, 0)$ to $(mn + j, n)$ that do not touch the line $x = my + j$ except at the end. In addition, if the maximum queue length does not exceed k, then the set of paths is further restricted not to cross the line $x = my + j - k - 1$, with the convention that the queue length does not include the one in service. The number of such paths is $|(\mathbf{b}, \mathbf{a}; 1)|$, and therefore the expression for $g(j, k, mn + j, m)$ follows. Thus the proof is complete.

Using some probabilistic reasoning, we are able to write

$$G(j, k, mn + j, m; t)$$

$$= g(j, k, mn + j, m) \int_0^t e^{-(\lambda_1 + \lambda_0)x}$$

$$\times \frac{(\lambda_1 + \lambda_0)x^{n(m+1)+j-1}}{(n(m + 1) + j - 1)!} (\lambda_1 + \lambda_0)\, dx. \qquad (5.32)$$

While our result is on a two-boundary situation, only the one-boundary case is treated by Takács [54]. Enns [13] has derived the expression for the generating function of the Laplace–Stieltjes transform of $G(j, k, u, l; t)$ in $M/G/1$ (i.e., service times have an arbitrary distribution). Notice that the number of departures (i.e., $mn + j$) is uniquely determined by the number of arriving batches (i.e., n). Therefore we could have used the latter as a parameter in place of the former. Furthermore, we may remark that by simply following the above argument the result can be readily modified if we introduce the change that the customers are served in batches instead of individually [36]. This is given as an exercise.

Next let us consider the model [36] which is the same as the above except that the customers arrive in batches of size u_i in accordance with a Poisson process of density $\lambda_i, i = 1, \ldots, k$. Let $G(j; a_1, \ldots, a_k; t)$ be the probability that a busy period, initiated by $j \geq 1$ customers, consists of a_i arriving batches of size u_i, $i = 1, \ldots, k$, and has length $\leq t$. Let $g(j; a_1, \ldots, a_k) = G(j; a_1, \ldots, a_k; \infty)$.

THEOREM 3

$$g(j; a_1, \ldots, a_k)$$

$$= \frac{j}{j + \sum_{i=1}^{k} (u_i + 1)a_i} \binom{j + \sum_{i=1}^{k} (u_i + 1)a_i}{a_1, \ldots, a_k}$$

$$\times \left(\frac{\lambda_0}{\lambda}\right)^{\sum_{i=1}^{k} u_i a_i + j} \prod_{i=1}^{k} \left(\frac{\lambda_i}{\lambda}\right)^{a_i}, \qquad (5.33)$$

where $\lambda = \lambda_0 + \lambda_1 + \cdots + \lambda_k$.

Proof

Suppose that at any instant the number of arriving batches of size u_i is x_i, $i = 1, \ldots, k$, and that the number of departing batches is x_0. Then the busy period continues as long as $x_0 < \sum_{i=1}^{k} u_i x_i + j$ and ends when for the first time $x_0 = \sum_{i=1}^{k} u_i x_i + j$. It consists of a_i arriving batches of size u_i, $i = 1, \ldots, k$, if and only if

(i) $x_0 < \sum_{i=1}^{k} u_i x_i + j$ when $0 \le x_i \le a_i$, $i = 1, \ldots, k$, and at least for some i, $0 \le x_i < a_i$, and

(ii) $x_0 = \sum_{i=1}^{k} u_i x_i + j$ when $x_i = a_i$, $i = 1, \ldots, k$.

This situation corresponds to the set of lattice paths from the origin to the point $(\sum_{i=1}^{k} u_i a_i + j, a_1, \ldots, a_k)$ that do not touch the hyperplane $x_0 = \sum_{i=1}^{k} u_i x_i + j$ except at the end. The number of such paths was given earlier.

The rest of the argument is routinely extended to this case from that of the previous one. This completes the proof.

The expression for $G(j; a_1, \ldots, a_k; t)$ is obtained from (5.33) as

$$G(j; a_1, \ldots, a_k; t)$$

$$= g(j; a_1, \ldots, a_k) \int_0^t e^{-\lambda y} \frac{(\lambda y)^{\sum_{i=1}^{k} (u_i + 1)a_i + j - 1}}{(\sum_{i=1}^{k} (u_i + 1)a_i + j - 1)!} \lambda \, dy. \qquad (5.34)$$

Lastly, we consider the queueing model as follows [37]. The inter-arrival times are i.i.d. random variables with distribution function F. The customers are served in batches of size m at a single counter. The departure times form a Poisson process with density μ and are independent of arrival times. In this model the server is idle (not busy) if and only if the queue size is smaller than m for the first time.

The busy period starts when there are m customers in the system. It consists of serving n batches if and only if there are $m(n-1)+j$, $j=0,1,\ldots,m-1$, arrivals and n departing batches. Let $G(n,m;j;t)$ $(n=1,2,\ldots;j=0,1,\ldots,m-1)$ denote the probability that (a) the busy period consists of serving n batches each consisting of m customers, (b) its length is $\leq t$, and (c) j customers are left in the system at the end of the busy period.

THEOREM 4

$$G(n,m;j;t) = \begin{cases} \mu \displaystyle\int_0^x e^{-\mu v}(1-F(v))\,dv \\ \qquad \text{when} \quad n=1, \quad r=1, \quad \text{and} \quad j=0, \\[2mm] \dfrac{\mu^n}{(n-1)!}\displaystyle\sum_{k=1}^n \binom{n-1}{k-1}\dfrac{m(k-1)+j}{m(n-1)+j} \\[2mm] \qquad \times \displaystyle\iint_R e^{-\mu(u+v)}u^{n-k}v^{k-1}(1-F(v))\,dF_{m(n-1)+j}(u)\,dv, \end{cases} \tag{5.35}$$

where $R = \{(u,v): u+v \leq t,\, u \geq 0,\, v \geq 0\}$ and F_k denotes the kth iterated convolution of F with itself.

Proof

The first part is obvious and is left out. Suppose the customers arrive at the instants $T_0, T_1, \ldots$, and let $X_n = T_n - T_{n-1}, n = 1, 2, \ldots$, which is the nth interarrival time. By our assumption the X_ns are i.i.d. random variables with $P[X_n \leq x] = F(x)$ for every n. Assume that the busy period, starting at $t = 0$ (i.e., $T_m = 0$), consists of n services each consisting of m customers. Let $y_i, i = 1, \ldots, m(n-1)+j+1$, denote the number of batches departing during the $(m+i-1)$th interarrival time. Then

$$\sum_{i=1}^{r} Y_i < \left[\frac{r-1}{m}\right] + 1, \qquad r = 1, \ldots, m(n-1)+j,$$

and

$$\sum_{i=1}^{m(n-1)+j+1} Y_i = n.$$

In this case the length of the busy period is given by $X_m + \cdots + X_{mn+j-1} + Z$, where Z is the time between the $(mn + j)$th arrival and the departure of the nth batch. Thus we have

$G(n, m; j; t)$

$$
= P\left(\sum_{i=1}^{r} Y_i < \left[\frac{r-1}{m} \right] + 1 \quad \text{for} \quad r = 1, \ldots, m(n-1) + j, \right.
$$

$$
\sum_{i=1}^{m(n-1)+j+1} Y_i = n, X_m + \cdots + X_{mn+j-1} + Z \le t \Big)
$$

$$
= \sum_{k=1}^{n} P\left(\sum_{i=1}^{r} Y_i < \left[\frac{r-1}{m} \right] + 1 \quad \text{for} \quad r = 1, \ldots, m(n-1) + j, \right.
$$

$$
\sum_{i=1}^{m(n-1)+j} Y_i = n - k, Y_{m(n-1)+j+1} = k,
$$

$$
X_m + \cdots + X_{mn+j-1} + Z \le t \Big). \tag{5.36}
$$

If $Y_{m(n-1)+j+1} = k$, then Z is independent of $Y_1, \ldots, Y_{m(n-1)+j}$ and $X_1, \ldots, X_{mn+j-1}$, and we have for $x \ge 0$ that

$$
P(Z \le x, X_{m(n-1)+j+1} = k) = \int_0^x e^{-\mu v} \frac{(\mu v)^{k-1}}{(k-1)!} (1 - F(v)) \mu \, dv. \tag{5.37}
$$

Furthermore,

$$
P(X_m + \cdots + X_{mn+j-1} \le u) = F_{m(n-1)+j}(u) \tag{5.38}
$$

and

$$
P\left(\sum_{i=1}^{m(n-1)+j} Y_i = n - k \,|\, X_m + \cdots + X_{mn+j-1} = u \right) = e^{-\mu u} \frac{(\mu u)^{n-k}}{(n-k)!}.
$$

$$
\tag{5.39}
$$

Under the condition $X_m + \cdots + X_{mn+j-1} = u$, $Y_1, \ldots, Y_{m(n-1)+j}$ are exchangeable and therefore

$$
P\left(\sum_{i=1}^{r} Y_i < \left[\frac{r-1}{m} \right] + 1 \quad \text{for} \quad r = 1, \ldots, m(n-1) + j \right|
$$

$$
\sum_{i=1}^{m(n-1)+j} Y_i = n - k, \sum_{i=m}^{mn+j-1} X_i = u \Big)
$$

$$
= 1 - \frac{m(n-k)}{m(n-1)+j} = \frac{m(k-1)+j}{m(n-1)+j} \tag{5.40}
$$

by (3.7).

On combining (5.36), (5.37), (5.38), (5.39), and (5.40) we get

$G(n, m; j; t)$

$$= \sum_{k=1}^{n} \frac{m(k-1)+j}{m(n-1)+j} \left\{ \left(\int_{0}^{t} e^{-\mu n} \frac{(\mu u)^{n-k}}{(n-k)!} \, dF_{m(n-1)+j}(u) \right) \right.$$

$$\left. * \left(\int_{0}^{t} e^{-\mu v} \frac{(\mu v)^{k-1}}{(k-1)!} (1 - F(v)) \mu \, dv \right) \right\}, \qquad (5.41)$$

where $*$ is the symbol of convolution. The proof is complete when we observe that (5.41) simplifies to the second part of (5.35).

Another model of interest [37], where (3.7) can be applied, is $M/G/1$ with the following conditions: the input of customers in batches of size m is a Poisson process of density λ; the service times are i.i.d. random variables with distribution function F, which are independent of arrival times. Letting $G(n, m; t)$ denote the probability that the length of the busy period, which consists of serving mn customers, is $\leq t$, we have the next theorem.

THEOREM 5

$$G(n, m; t) = \frac{\lambda^{n-1}}{n!} \int_{0}^{t} e^{-\lambda u} u^{n-1} \, dF_{nm}(u), \qquad n = 1, 2, \ldots. \quad (5.42)$$

The proof is left as an exercise.

4. Trees and Search Codes

It is not unusual to think that lattice paths should be in some way associated with trees. In fact it is known ([38, p. 6]) that the number of labeled trees with n vertices such that the vertex labeled i has degree $d_i, i = 1, \ldots, n$, is

$$\binom{n-2}{d_1 - 1, \ldots, d_{n-1} - 1}, \qquad (5.43)$$

which is equal to the number of lattice paths from the origin to $(d_1 - 1, \ldots, d_n - 1)$. Other recent results [4, 29, 30, 49] readily provide further evidence in this direction. In the sequel we present some of these results, show their interconnections, and ultimately come to the well-known result due to Cayley that the number of labeled trees with n vertices is n^{n-2}. Later we define search codes and point out their relations to trees.

Let $T = (V, E, v, \alpha)$ be a planted plane (in short p.p.) tree, where V is the vertex set, E the edge set (a set of two sets of V), v the distinguished vertex, called the root whose degree is one, and α the order relation on V, that possesses the following properties.

(i)   For $x, y \in V$ if $\rho(x) < \rho(y)$, then $x\alpha y$, where $\rho(x)$ is the length of the path from v to x.

(ii) If $\{r, s\}, \{x, y\} \in E$, $\rho(r) = \rho(x) = \rho(s) - 1 = \rho(y) - 1$, and $r\alpha x$, then $s\alpha y$.

Two p.p. trees (V, E_1, v_1, α_1) and (V, E_2, v_2, α_2) are isomorphic if there exists a permutation ϕ of V such that

$$\phi(v_1) = v_2, \qquad E_2 = \{\{\phi(x), \phi(y)\} : \{x, y\} \in E_1\}, \qquad (5.44)$$

and $x\alpha_1 y$ if and only if $\phi(x)\alpha_2 \phi(y)$. One can easily draw a picture of a p.p. tree by arranging the vertices in levels so that vertex x is in level $\rho(x)$ and then arranging the vertices in each level from left to right according to the order relation α. Without the order relation the tree (V, E, v) is a planted tree. We remark that two p.p. trees are isomorphic when they have the same form of diagram in the plane. The original definition of a p.p. tree given by Harary et al. [21] is different from the above due to Klarner [30].

Consider $r = (r_1, \ldots, r_m)$ and $q = (q_1, \ldots, q_m)$, where the r_is are nonnegative integers and the q_is are positive integers such that

(a) $q_1 + \cdots + q_i - i \geq r_1 + \cdots + r_i$ $\quad (i = 1, \ldots, m - 1)$, and

(b) $q_1 + \cdots + q_m - m + 1 = r_1 + \cdots + r_m$.

Denote by $P(V, q, r)$ the set of p.p. trees with the following two conditions:

(i)   There are exactly m vertices of degree greater than one (we call such a vertex a branch point) such that the ith one of these vertices (as ordered by α) has exactly degree $q_i + 1$, $i = 1, \ldots, m$.

(ii) Let x_i be the total number of vertices of degree one between the first and $(i + 1)$st branch point, $i = 1, \ldots, m - 1$, and x_m the total number of vertices of degree one excluding the root. The condition is that we have $x_i \geq r_1 + \cdots + r_i$ $(i = 1, \ldots, m - 1)$ and $x_m = r_1 + \cdots + r_m$.

Level

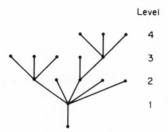

4

3

2

1

Figure 1. A tree with $q = (5, 3, 2, 3)$, $x = (0, 1, 7, 10)$.

It is clear from the definition that

$$|V| = \sum_{i=1}^{m} q_i + 2 = \sum_{j=1}^{m} r_j + m + 1.$$

Also we remark that each p.p. tree $T = (V, E, v, \alpha)$ determines a vector $(x_1, \ldots, x_m)$ and that any two isomorphic trees have the same vector but not conversely. We call $(x_1, \ldots, x_m)$ corresponding to T its characteristic vector (see Fig. 1).

Denote by IP the set of classes of isomorphic trees in any set of trees P. Recalling the notation $(\mathbf{b}, \mathbf{a})$ from Section 2 of Chapter 2, we find $|IP(V, q, r)|$ in the following theorem.

THEOREM 6

$$|IP(V, q, r)| = |(\mathbf{b}, \mathbf{a})|, \tag{5.45}$$

where

$$b_i = \sum_{j=1}^{i} r_j \quad \text{and} \quad a_i = \sum_{j=1}^{i} q_j - i, \quad i = 1, \ldots, m - 1.$$

Proof

To prove the theorem we have to show that there exists a bijective mapping between $IP(V, q, r)$ and $(\mathbf{b}, \mathbf{a})$, which represents the set of lattice paths that dominate $\mathbf{b} = (b_1, \ldots, b_{m-1})$ and are dominated by $\mathbf{a} = (a_1, \ldots, a_{m-1})$.

Let T, with characteristic vector $(x_1, \ldots, x_m)$, be a representative tree from an isomorphic class $IP(V, q, r)$. By definition the ith branch point has degree $q_i + 1$, which implies that

$$x_i \le \sum_{j=1}^{i} q_j - i, \quad i = 1, \ldots, m - 1.$$

Moreover,

$$0 \le x_1 \le \cdots \le x_{m-1} \quad \text{and} \quad \sum_{j=1}^{i} r_j \le x_j.$$

Because of the earlier remark on the representation of p.p. trees it is obvious that to representative trees of two distinct isomorphic classes there correspond two different vectors belonging to $(\mathbf{b}, \mathbf{a})$.

The inverse mapping is quite natural. Given a vector $(x_1, \ldots, x_{m-1}) \in$ $(\mathbf{b}, \mathbf{a})$, we construct a representative tree of a class in $IP(V, q, r)$ as follows. The first branch point of such a tree is the second vertex. Draw q_1 outgoing edges from this branch point (i.e., one edge joining this vertex at level 1 and q_1 vertices at level 2). We start at level 2 and count vertices from a lower level to the next higher level and from left to right within a level. Adopting this counting procedure, assign the second branch point of vertex $x_1 + 1$, at which point draw q_2 outgoing edges (which are at the next higher level), and in general assign the $(i + 1)$st $(i = 1, \ldots, m - 1)$ branch point at the $(x_i + i)$th vertex, at which point draw q_{i+1} outgoing edges. This is always possible because

$$x_i + i \le \sum_{j=1}^{i} q_j \quad (i = 1, \ldots, m - 1).$$

The number of vertices of degree one, excluding the root, is $q_1 + \cdots + q_m - m + 1 = r_1 + \cdots + r_m$. Thus the tree so constructed is in $IP(V, q, r)$. This completes the proof.

Since $|(\mathbf{b}, \mathbf{a})|$ has an explicit expression (see Theorem 1, Chapter 2), so does $|IP(V, q, r)|$.

Next we consider another set of trees, denoted by $P^*(V, k)$, $k = (k_1, \ldots, k_n)$, which represents the set of p.p. trees such that exactly $k_i (\ge 0)$ vertices have degree $i + 1$, $i = 1, \ldots, n$. Letting $\sum$ represent $\sum_1^n$, we have

$$|V| = \sum ik_i + 2. \tag{5.46}$$

The following theorem gives $|IP^*(V, k)|$ explicitly, again by using the lattice path correspondence. (Also see Tutte [59].)

THEOREM 7

$$|IP^*(V, k)| = \frac{1}{n+1} \binom{n+1}{k_1, \ldots, k_n}, \tag{5.47}$$

where $n = \sum ik_i$. (If we consider partitions of n in nondecreasing parts, the frequencies of unequal parts become k_is.)

Proof

Let $B^*(k)$ denote the set of $(n + 1)$-dimensional lattice paths from the origin to $(x_0, x_1, \ldots, x_n) = (\sum (i - 1)k_i + 1, k_1, \ldots, k_n)$ which lie beneath the hyperplane $x_0 = \sum (i - 1)x_i$ except at the origin. Since we already know that $|B^*(k)|$ equals the right-hand side of (5.47), it is only necessary to establish a $1 : 1$ correspondence between $IP^*(V, k)$ and $B^*(k)$.

Take a representative tree from an isomorphic class in $IP^*(V, k)$. From (5.46) it follows that excluding the root there are exactly $\sum (i - 1) \times k_i + 1$ vertices having degree one. Let v_j be the $(\sum ik_i - j + 3)$th vertex according to the order relation α on V. Starting from the last vertex, represent v_j $(j = 1, \ldots, \sum ik_i)$ by a unit on the X_i-axis if its degree is $i + 1$ $(i = 0, 1, \ldots, n)$.

The construction is illustrated in Fig. 2. Let β_i^j be the number of units on the X_i-axis up to and including the jth step. We need to show that $\beta_0^j > \sum (i - 1)\beta_i^j$. Note that the unit step on the X_i-axis, $i \neq 0$, corresponds to a branch point of degree $i + 1$, and associated with this step there should be i steps preceding it. Therefore if the jth step is not on the X_0 axis, the total number of steps up to and including the jth step is at least $\sum i\beta_i^j + 1$; i.e., $\sum_{i=0}^n \beta_i^j \geq \sum i\beta_i^j + 1$. Consequently $\beta_0^j > \sum (i - 1)\beta_i^j$. A similar argument applies to the case in which the jth step is on the X_0-axis.

From the above illustration the inverse mapping is simple to construct. Paths corresponding to different members of $IP^*(V, k)$ are different. Thus we have completed the proof.

Tree   Path

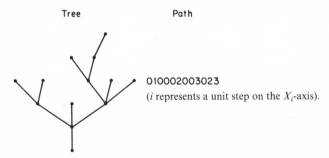

010002003023
(i represents a unit step on the X_i-axis).

Figure 2

Interestingly enough, these two theorems lead to the following combinatorial identity a direct proof of which is not obvious. In the first theorem let $r = 0$ mean $r_i = 0$ for all i and let there be exactly k_i qs, each equal to i, $i = 1, \ldots, n$. Under these conditions we get the identity

$$\sum |IP(V, q, 0)| = |IP^*(V, k)|, \qquad (5.48)$$

where the summation is over all distinct permutations of $(q_1, \ldots, q_m)$. The proof of the identity even when using the interpretation of lattice paths does not seem to be straightforward.

Let $P(V, n)$ be the set of p.p. trees with $n + 1$ edges (i.e., with $n + 2$ vertices). We determine $|IP(V, n)|$, again using the path correspondence method due to Harris [22] (for an alternate proof see [12].)

THEOREM 8

$$|IP(V, n)| = \frac{1}{2n + 1} \binom{2n + 1}{n}. \qquad (5.49)$$

Proof

Recall that the number on the right-hand side equals the number of paths from the origin to $(n + 1, n)$ that do not touch the line $x = y$ except at the origin [see (1.4) of Chapter 1]. Therefore, it suffices to establish a bijectivity between $IP(V, n)$ and the above set of paths. Since there is an obvious correspondence between paths and walks (Chapter 4), to each representative tree in $IP(V, n)$ we associate a random walk as follows.

Let each vertex be assigned the number of the level to which it belongs. Starting from the root, we move to the vertex at level 1. The walk corresponding to this step is the movement from 0 to 1. Suppose that at a given stage we are at vertex v with level number k. In the next stage we move from left to right to a vertex at the next higher level, which is joined to v through an edge. In this case the walk correspondingly moves from k to $k + 1$. If either all vertices at level $k + 1$ are exhausted or there is no vertex at level $k + 1$ joined to v (i.e., the degree of v is one), we move to the vertex joined to it at level $k - 1$, which corresponds to the transition of the walk from k to $k - 1$. This process ends when we reach the vertex at level 1. Thus at any stage in the random walk we move a unit step either to the left or to the right. It can be easily seen that we traverse twice on each edge, once going up and another time coming down, except for the one joining the root which is traversed

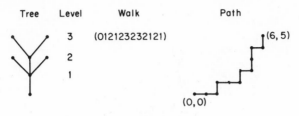

| Tree | Level | Walk | Path |

Figure 3

only once. In other words the walk described by the sequence of level numbers starts from 0 and ends at 1 in $2n + 1$ steps without reaching 0 at any stage. The illustration (see Fig. 3) will help to clarify the construction procedure. The set of paths corresponding to $IP(V, n)$ is clearly that described at the beginning of the proof. That the mapping is bijective is evident from Fig. 3. This completes the proof.

The two previous theorems give rise to another interesting combinatorial identity, which is

$$\sum \frac{1}{n+1} \binom{n+1}{k_1, \ldots, k_n} = \frac{1}{2n+1} \binom{2n+1}{n}, \qquad (5.50)$$

the summation being over all possible $k = (k_1, \ldots, k_n)$ that are solutions of $\sum i k_i = n$.

Another significant consequence of (5.47) is an alternative derivation of Cayley's result on labeled trees. The relation between (5.47) and the number of labeled trees with $n + 1$ vertices is given, in a way, by Riordan [49], in an indirect manner by means of another problem. However our approach is direct and uses a bijective mapping constructed for the specific purpose.

For any n consider all possible solutions k of $\sum i k_i = n$. For a given k we form a nondecreasing sequence of n positive integers corresponding to every member of $IP^*(V, k)$ when we assign j $(j = 1, \ldots, \sum k_i)$ to all outgoing edges from the jth branch point. The association scheme is illustrated in Fig. 4. Take all possible distinct permutations of the sequence, the number being equal to

$$\frac{n!}{\prod_{j=1}^{n} (j!)^{k_j}}.$$

For any fixed n consider the set of all such permutations for all solutions k and let this set be denoted by $P^*(n)$.

Member of $IP^*(V, k)$ with
$k = (3, 2, 1, 0, 0, 0, 0, 0, 0, 0)$.

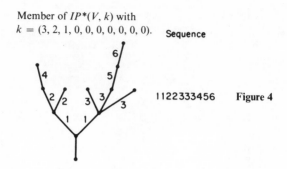

Sequence

1122333456   **Figure 4**

THEOREM 9

There is a 1 : 1 correspondence between $P^*(n)$ and the set of labeled trees with $n + 1$ vertices.

Proof

Let $(p_1, \ldots, p_n)$ be a particular permutation in $P^*(n)$. Considering the member of $IP^*(V, k)$ corresponding to $(p_1, \ldots, p_n)$, we construct a labeled tree as follows. Label the vertex at level one (which is the first branch point) with $n + 1$ and put label i at the end vertex of an outgoing edge marked p_i such that the vertex labeled i lies to the left of vertex labeled j if $p_i = p_j$, $i < j$. The tree so constructed is a labeled tree with $n + 1$ vertices. For example, corresponding to permutation (3211433625) of (1122333456) is the labeled tree shown in Fig. 5. Obviously, any two labeled trees from two different permutations that arise from a given member of $IP^*(V, k)$ are different. It can be seen that two trees corresponding to two different members of $IP^*(V, k)$ are also different.

The inverse mapping is as follows. Given a labeled tree, we put the tree as a p.p. tree, arranged in different levels so that the vertex with

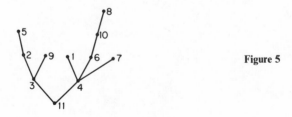

Figure 5

label $n + 1$ becomes the first branch point, and at each level labels of vertices which are endpoints of outgoing edges from any branch point in the next lower level form an increasing sequence from left to right. Evidently it can be done in only one way. Use the ordering relation α to order the vertices and assign j to all outgoing edges from the jth branch point. Then we write a sequence of numbers such that the ith number is the number assigned to the outgoing edge whose end vertex has label i. Clearly, the sequence of numbers is a permutation belonging to $P^*(n)$. This completes the proof.

From the above theorem it follows that the number of labeled trees with $n + 1$ vertices is given by

$$\sum \frac{1}{n+1} \binom{n+1}{k_1, \ldots, k_n} \frac{n!}{\prod_{j=1}^{n} (j!)^{k_j}}, \qquad (5.51)$$

where the summation is over all solutions $(k_1, \ldots, k_n)$ of $\sum ik_i = n$. It can be proved [48, 49] that (5.51) simplifies to $(n + 1)^{n-1}$, which is Cayley's result. For an equivalent parking space problem see [49].

As remarked earlier, the $1:1$ correspondence of the last theorem, which is made explicit here, exists tacitly in [49].

Next we consider a certain set of isomorphic planted edge-chromatic trees. Letting $K = \{1, \ldots, d + 1\}$, a planted edge-chromatic (in short PEC) tree on K, denoted by $(V, E, v; f, \gamma)$, is defined to be a planted tree, where

(i) $f: E \to K$ satisfies that the rooted edge is mapped to one and adjacent edges are mapped to distinct elements of K (in other words, we color edges with $d + 1$ colors such that the rooted edge is assigned color 1 and no two adjacent edges have the same color); and

(ii) γ is an order relation on V, possessing the properties of the order relation and satisfying the following: For

$$\{r, x\}, \{r, y\} \in E \qquad \text{and} \qquad \rho(r) = \rho(x) - 1 = \rho(y) - 1,$$

$$x\gamma y \qquad \text{if} \quad f\{r, x\} < f\{r, y\}.$$

Two PEC trees $T_1 = (V, E_1, v_1; f_1, \gamma_1)$ and $T_2 = (V, E_2, v_2; f_2, \gamma_2)$ are isomorphic if there exists a permutation ϕ on V such that in addition to the conditions (5.44) we have $f_1\{x, y\} = f_2\{\phi(x), \phi(y)\}$. It is easy to check that $x\gamma_1 y$ in T_1 if and only if $\phi(x)\gamma_2 \phi(y)$ in T_2. Let $C(V, K)$ denote

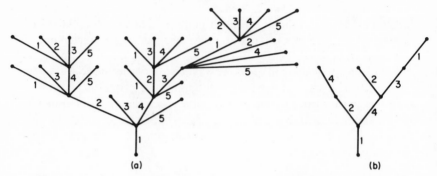

Figure 6.   (a) A member of $IP(V^*, q^*, 0)$; (b) corresponding member of $IC(V, K)$.

the set of PEC trees on K having vertex set V. By using the generating function Carlitz [3] enumerated the set $IC(V, k)$, which is shown by Klarner [29] to have $1:1$ correspondence with $IP(V^*, q^*, 0)$, where $|V^*| = dm + 2$ and $q_i^* = d$ if and only if $|V| = m + 1$ for $i = 1, \ldots, m$. We explain the correspondence and illustrate it below.

For a given number T of $IP(V^*, q^*, 0)$ construct the unique subtree T' of T which has as its vertex set the root and all vertices of degree $d + 1$ in T. Since T has exactly m vertices of degree $d + 1$, T' has exactly $m + 1$ vertices. We color T by first assigning color 1 to the rooted edge and moving upward in an inductive manner. Suppose the edge joining x and the vertex below it has been assigned color i and suppose $c_1 < \cdots < c_d$ are the colors in $K/\{i\}$. Then the outgoing d edges from x (if it is a branch point) are colored $c_1, \ldots, c_d$ from left to right. This coloring of T is also a coloring of T', where the edges of T' have the same colors as those of the original T. Clearly T' is a member of $IC(V, K)$ with $|V| = m + 1$. See Fig. 6 as an illustration of the construction. Here we have taken $d = 4$, $m = 7$. It can be seen that the construction is a bijective mapping between $IC(V, K)$ and $IP(V^*, q^*, 0)$.

It is possible to define a generalized PEC tree and find isomorphic classes that correspond to $IP(V, q, r)$ and $IP^*(V, k)$ [5]. This is left as an exercise for the reader.

It is but natural to expect from the connection between paths and trees that Lagrange's expansion should play a role in certain types of tree enumeration. In fact this remark is corroborated by Good's work [17, 18].

Having discussed trees, we move on to introduce search codes, originally considered by Rényi [47] in the theory of search. We need the following set of definitions.

(1) Any finite sequence of nonnegative integers is a *code word*. We denote by Z the set of all code words. It is convenient to consider the empty sequence e to be a code word.

(2) A code word b is a *prefix* of a code word c if there exists a code word d such that $c = bd$, where bd is the concatenation of the code words b and d.

(3) Any finite set C of different code words is a *code*. The empty set $\varnothing$ is considered to be a code and is called the *empty code*. The code consisting of the empty code word e is called the *trivial code*.

If C is a code and a any code word, denote by C_a the set of all code words $b \in Z$ such that $ab \in C$.

(4) A code C is *branched* if one of the following occurs:

  (i) C is the empty code $\varnothing$;

  (ii) C is the trivial code $\{e\}$:

  (iii) C does not contain e and there exists an integer $\delta(C) \geq 1$ such that for the code word consisting of the single letter $k, k = 0, 1, \ldots,$ the code C_k is empty or nonempty according as $k \geq \delta(C)$ or $k < \delta(C)$. (In other words every nonnegative integer $k < \delta(C)$ must be a prefix to some code word in C, and there is no code word in C with k as a prefix if $k \geq \delta(C)$.)

(5) C is a *pseudo-search code* if C_a is branched for every $a \in Z$.

(6) For a pseudo-search code C any $a \in Z$ for which $\delta(C_a) \geq 1$ is a *branch point* of C, and $\delta(C_a)$ is the *branching number* of the branch point a. It is convenient to say that the branch point a has $\delta(C_a)$ branches.

(7) C is a *search code* if $\delta(C_a) \geq 2$ for every branch point a of C.

(8) A pseudo-search code is *regular* of degree q (≥ 1) if the branching number of each branch point is q.

To clarify the above definitions, consider the codes

$$C^1 = \{0, 10, 11, 2\}, \qquad C^2 = \{0, 11, 2\}, \qquad C^3 = \{0, 21\}.$$

C^1 is a pseudo-search code, C^2 is branched but is not a pseudo-search code (since C_1^2 is not branched), and C^3 is not branched (since $C_1^3 = \varnothing$, but $C_2^3 = \{1\}$).

The words "branched," "branch point," and "branches" are borrowed from the terminology of trees, and indeed a pseudo-search

Figure 7

code is in correspondence with a tree by a particular construction procedure. For example, the tree corresponding to C^1 is given in Fig. 7. Another instance is shown in Fig. 8, in which the correspondence with the path is also demonstrated. The tree in Fig. 8 happens to be a member of $IP(V, q, 0)$, having $q_1 = 3$, $q_2 = 1$, $q_3 = 4$, $q_4 = 1$, and its correspondence with the path was explained in the proof of Theorem 6. Note that the vertices of degree one from left to right correspond to the code words when these are lexicographically arranged.

Let C be a pseudo-search code and $B(C)$ the set of its branch points. Let $|B(C)| = m$. Arrange the code words in $B(C)$ in lexicographic order and label these 1, 2, ..., m in that order. Denote by $S(q_1, ..., q_m)$ the set of all pseudo-search codes such that the ith branch point has q_i, $i = 1, ..., m$, as its branching number. The above illustration suggests that one can readily establish a bijective mapping between $S(q_1, ..., q_m)$ and $IP(V, q, 0)$. In fact we can write a set of pseudo-search codes corresponding to $IP(V, q, r)$ (see [4]).

As a special case we count the regular search codes with m branch points, each having u as its branching number, the expressions being

$$\frac{1}{mu + 1}\binom{mu + 1}{m},$$

which was derived by Rényi [47] by using the generating function.

By no means do we claim that the references and results presented in this section give a complete picture of the relations between paths and

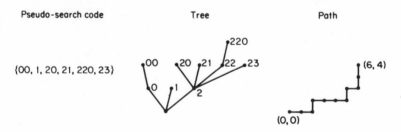

Figure 8

trees. Some of them appear in the exercises. Mullin's work [39–41] on triangular maps, which are related to p.p. trees, has not been included. The interested reader may also refer to [10, 60].

5. Concluding Remarks

We conclude this chapter by pointing out applications that stand apart from the rest. Because of the technical nature of the first problem, we avoid describing it in detail, but arrive at the combinatorial aspect by glossing over it fairly quickly. In Section 2 we became familiar with a sequential sampling plan on a simple random walk, viz., the negative binomial sampling plan, for the estimation of p. A general class of sequential sampling plan on a simple random walk as first introduced by Girshick *et al.*, [16] and later studied by DeGroot [11] is called simple sampling plans (briefly s.s.p.s). A typical s.s.p. is illustrated in Fig. 9. The points marked with x and $\odot$ are, respectively, boundary points (briefly b.p.s) and accessible points of the plan. A sample in the plan is a lattice path from the origin to a boundary point through accessible points. We say that a s.s.p. has size n if $\max_B\{x + y : (x, y) \in B\} = n$, where B is the set of b.p.s. In our example $n = 5$. A combinatorial question is to count the s.s.p.s of size n.

An important property of a s.s.p. of size n is that a sampling plan always has $n + 1$ b.p.s if and only if it is simple. These b.p.s may be ordered as suggested in Fig. 9. For the purpose of counting we give a characterization (for details see [43]) of s.s.p.s of size n in terms of vectors of $n + 1$ elements corresponding to $n + 1$ b.p.s, so that if the ith b.p. is (x_i, y_i), the ith element of the vector is $n - (x_i + y_i)$, $i = 1, \ldots, n + 1$. In Fig. 9 the vector representing the s.s.p. is $(4, 1, 0, 0, 2, 2)$.

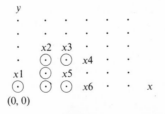

Figure 9

In general, it is proved [43] that a vector $(a_1, \ldots, a_{n+1})$ which corresponds to a s.s.p. of size n satisfies the following:

    (i)   There exists an integer i, $1 \leq i \leq n$, such that $a_i = a_{i+1} = 0$.

    (ii) $a_1 \geq \cdots \geq a_{k-1} > 0$ and $0 \leq a_{k+2} \leq \cdots \leq a_{n+1}$, where k is the smallest integer such that $a_k = a_{k+1} = 0$ $(1 \leq k \leq n)$.

    (iii) Let $(\alpha_1, \ldots, \alpha_{n-1})$ be a nondecreasing ordering of $a_1, \ldots, a_{k-1}$ and $a_{k+2}, \ldots, a_{n+1}$ subject to the convention that if any of $a_1, \ldots, a_{k-1}$ is equal to any of $a_{k+2}, \ldots, a_{n+1}$, the subscript of α corresponding to the first set is smaller than that of the second. Furthermore, let

$$\beta_i = \begin{cases} 2\alpha_i - 1 & \text{if } \alpha_i \in \{a_1, \ldots, a_{k-1}\}, \\ 2\alpha_i & \text{if } \alpha_i \in \{a_{k+2}, \ldots, a_{n-1}\} \end{cases}$$

for $i = 1, \ldots, n-1$. Then

$$0 \leq \beta_1 \leq \cdots \leq \beta_{n-1} \quad \text{and} \quad \beta_i \leq 2i \quad \text{for} \quad i = 1, \ldots, n-1.$$

$$(5.52)$$

(For characterizations of these plans along with others see Kagan *et al.*, [28, Chap. 12].)

Thus the number of s.s.p.s of size n is equal to the number of vectors $(\beta_1, \ldots, \beta_{n-1})$ satisfying (5.52), the expression for which is $n^{-1}\binom{3n}{n-1}$ by (1.11) and Section 6, Chapter 1. The usefulness of this combinatorial result in statistical applications is not yet established. Discussion of symmetric s.s.p.s is included in [43].

Finally, we mention a bizarre application of (2.1) due to Kryscio and Severo [31]. Many epidemic models are special cases of the so-called multidimensional right shift processes (in short MRSP), the Kolmogorov forward equations of which should be solved for any relevant probability statement. It turns out that in estimating the unknown parameters of a MRSP when the process is observed at a finite number of fixed time points only a subset of the Kolmogorov equations is to be solved. For computational purposes in solving these equations an upper bound of the cardinality of the subset is helpful and happens to be provided by (2.1).

Exercises

1. Show that [25] for $0 < p < 1$ and for any real μ such that $|(\mu + 1)q| < 1$, (5.3) is satisfied.

2. Check (5.4) and derive (5.5).

3. Find the third and fourth moments of the GNB distribution [25].

4. Verify (5.6) and find the mean and variance of the generalized Poisson distribution.

5. For what conditions does (5.10) define a probability distribution?

6. (a) Referring to (5.10), show that

$$\dfrac{\dfrac{r_1}{r_1 + \sum (\mu_i + 1)k_i} \left(\begin{matrix} r_1 + \sum (\mu_i + 1)k_i \\ k_1, \ldots, k_j \end{matrix}\right) \times \dfrac{r_2}{r_2 + \sum (\mu_i + 1)(n_i - k_i)} \left(\begin{matrix} r_2 + \sum (\mu_i + 1)(n_i - k_i) \\ n_1 - k_1, \ldots, n_j - k_j \end{matrix}\right)}{\dfrac{r_1 + r_2}{r_1 + r_2 + \sum (\mu_i + 1)n_i} \left(\begin{matrix} r_1 + r_2 + \sum (\mu_i + 1)n_i \\ n_1, \ldots, n_j \end{matrix}\right)}$$

for $k_i = 0, 1, \ldots, n_i, i = 1, \ldots, j$, defines a probability distribution. This is called a generalized hypergeometric distribution.

(b) Multisample Pólya urn scheme (see [14, p. 110]). From an urn containing b black balls and r red balls we draw a ball at random and then replace it along with an additional c balls of the same color. A new random drawing is made and the procedure is repeated for a particular sample. In this manner draw j independent samples of sizes $n_1, \ldots, n_j$. Let X_i be the number of black balls in the ith sample, $i = 1, \ldots, j$. Show that the joint distribution of $X_1, \ldots, X_j$ is a generalized hypergeometric distribution. Also prove that

$$E\left(\prod_{j=1}^{j} X_i(X_i - 1) \cdots (X_i - m_i + 1) \right)$$

$$= \dfrac{(\alpha + \gamma - 1)!(\alpha + \sum m_i - 1)!}{(\alpha - 1)!(\alpha + \gamma + \sum m_i - 1)!} \prod_{i=1}^{j} n_i(n_i - 1) \cdots (n_i - m_i + 1),$$

where $\alpha = b/c$ and $\gamma = r/c$.

7. Verify (5.25) and (5.26).

8. In the correlated random walk find MVU estimates of p_1 and p_2 for $r = 1$.

9. For general $\mu \geq 0$ find the first passage time distributions, find the corresponding means and variances, and obtain MVU estimates of p_1 and p_2.

10. Consider the two-coin tossing game with the following rules (let the number of heads and tails at any stage of the game be α and β, respectively):
(1) Continuing rule. Next toss coin $a + 1$ if $\alpha \equiv a \pmod 2$, $a = 0, 1$. (It is obvious that the game starts with coin 1.)

(2)   Stopping rule.   Stop tossing when for the first time $\alpha + 2\beta > r$.

Find the distribution of the duration of the game and discuss the MVU estimates of p_1 and p_2.

11.   Consider the two-coin tossing game $H_i(r)$, $i = 1, 2$, with the following rules:

(1)   Start with coin i, $i = 1, 2$.
(2)   Toss coin 1 and 2 alternately.
(3)   Stop tossing when for the first time the total number of heads exceeds the total number of tails by exactly r.

(a)   Show that the probability generating functions (p.g.f.s in short) of the duration of the game for $H_1(1)$ and $H_2(1)$ are, respectively,

$$\frac{1 + u - v - \sqrt{1 + u^2 + v^2 - 2u - 2v - 2uv}}{2q_2 s}$$

and

$$\frac{1 - u + v - \sqrt{1 + u^2 + v^2 - 2u - 2v - 2uv}}{2q_1 s},$$

where $u = p_1 q_2 s^2$ and $v = q_1 p_2 s^2$.

(b)   Obtain expressions of p.g.f.s of the duration of the game for $H_1(1)$, $H_2(1)$, and $H_1(2) = H_2(2)$ by proving the identity

$$(1 + u^2 + v^2 - 2u - 2v - 2uv)^{1/2}$$

$$= 1 - u - v - 2 \sum_{i=1}^{\infty} \sum_{k=1}^{\infty} \frac{\binom{j+k-1}{j}\binom{j+k-1}{k}}{(j+k-1)} u^j v^k.$$

(c)   Find the mean and variance of the duration of the game $H_i(r)$.

12.   In the coin tossing game of Exercise 11, we consider either of the following as a subsidiary sequence (s.s.):

(i)   an H with coin 1 followed by a T with coin 2;
(ii)   a T with coin 1 followed by an H with coin 2. A basic sequence (b.s.) is one without any s.s. Using the notions of b.s.s and s.s.s, show that the p.g.f. of the duration of the game for $H_i(2k)$ is

$$\sum_{n=0}^{\infty} \frac{k}{2n+k} \binom{2n+k}{n} p^{n+k} q^n \left(\frac{s^2}{1 - rs^2}\right)^{2n+k},$$

where $p = p_1 p_2$, $q = q_1 q_2$, and $r = p_1 q_2 + q_1 p_2$.

13. Prove (5.32).

14. In the model of Theorem 2 suppose customers are served in batches of size v by a single server. Letting $G^*(j, k, u; t)$ be the probability that a busy period initiated by $j \geq v$ customers has a maximum queue length $\leq k$ that consists of u arriving batches and has length $\leq t$, find an expression for $G^*(j, k, u; t)$.

15. Prove Theorem 5.

16. Consider the queueing model $M/M/c$ in which customers arrive according to a Poisson process of density λ, join a single queue, and are served by a service system consisting of c channels. The service times of the ith channel $i = 1, \ldots, c$, are i.i.d. random variables with distribution function

$$F(x) = \begin{cases} 1 - \exp(-\mu_i x) & \text{if } x \geq 0, \\ 0 & \text{otherwise,} \end{cases}$$

and are independent of service times and arrival times. The busy period is defined as the time interval during which all c channels are busy. Letting $G(j; k_1, \ldots, k_c; t)$ be the probability that the busy period initiated by j ($\geq c$) customers has length $\leq t$ and consists of serving k_i customers at channel i ($i = 1, \ldots, c$), show that

$$G(j; k_1, \ldots, k_c; t)$$
$$= \frac{j - c + 1}{m!} \lambda^m \prod_{i=1}^{c} \left(\frac{\mu_i^{k_i}}{k_i!} \right) \int_0^t e^{-(\lambda + \mu)y} y^{2m+j-c} \, dy,$$

where $\mu = \sum_{i=1}^{c} \mu_i$ and $m \, (= \sum_{i=1}^{c} k_i - j + c - 1)$ represents the number of arrivals during the busy period. (This problem was solved by the author jointly with B. R. Handa and J. L. Jain.)

17. Prove that the number of classes of isomorphic p.p. trees with $2n + 2$ vertices such that there are exactly n vertices of degree 3 and others of degree 1 is $|IP(V, n)|$.

18. By using Theorem 6 and Theorem 8, respectively, prove the following identities [30]:

(i) $\sum \binom{kn_1}{n_2} \binom{kn_2}{n_3} \cdots \binom{kn_{j-1}}{n_j} = \frac{1}{k(n+1)+1} \binom{k(n+1)+1}{n+1}$,

(ii) $\sum \binom{n_1 + n_2 - 1}{n_2} \binom{n_2 + n_3 - 1}{n_3} \cdots \binom{n_{j-1} + n_j - 1}{n_j}$
$$= \frac{1}{2n+1} \binom{2n+1}{n},$$

where the sum in either case extends over all compositions $(n_1, \ldots, n_j)$ of $n + 1$ into an unrestricted number of positive parts with $n_1 = 1$.

19. By two different ways of counting trees, prove the identity [50]

$$\sum \frac{1}{n + 1} \binom{n + 1}{k_1, \ldots, k_n} \prod_{i=1}^{n-1} \binom{k}{i}^{k_i} = \frac{1}{k(n + 1) + 1} \binom{k(n + 1) + 1}{n + 1}.$$

20. Prove that (5.51) equals $(n + 1)^{n-1}$.

21. Define a generalized PEC tree and find isomorphic classes which are in $1:1$ correspondence with $IP(V, q, r)$ and $IP^*(V, k)$.

22. Prove that $S(q_1, \ldots, q_m)$ and $IP(V, q, 0)$ are in $1:1$ correspondence.

References

1. Boswell, M. T., and Patil, G. P., Chance mechanisms generating the negative binomial distributions, in *Random Counts in Scientific Work* (G. P. Patil, ed.), Vol. 1: Random Counts in Models and Structures, pp. 3–22. Pennsylvania State Univ. Press, University Park, Pennsylvania, 1970.

2. Cacoullos, T., and Charalambides, Ch., M.V.U.E. for truncated discrete distributions, *Colloquia Mathematica Societatis János Bolyai* **9** (1972), 133–144. (European Meeting of Statisticians, 1972). North-Holland Publ., Amsterdam.

3. Carlitz, L., A note on the enumeration of line chromatic trees, *J. Combin. Theory* **6** (1969), 99–101.

4. Chorneyko, I. Z., and Mohanty, S. G., On the enumeration of pseudo-search codes, *Studia Sci. Math. Hungar.*, **7** (1972), 47–54.

5. Chorneyko, I. Z., and Mohanty, S. G., On the enumeration of certain sets of planted plane trees, *J. Combin. Theory* **18** (1975), 209–221.

6. Consul, P. C., and Jain, G. C., A generalization of the Poisson distributions, *Technometrics* **15** (1973), 791–799.

7. Consul, P. C., and Jain, G. C., On some interesting properties of the generalized Poisson distribution, *Biometrische Z.*, **15** (1973), 495–500.

8. Consul, P. C., and Shenton, L. R., Use of Lagrange expansion for generating discrete generalized probability distributions, *SIAM J. Appl. Math.* **23** (1972), 239–248.

9. Consul, P. C., and Shenton, L. R., Some interesting properties of Lagrangian distributions, *Comm. Statist.* **2** (1973), 263–272.

10. De Bruijn, N. G., and Morselt, B. J. M., A note on plane trees, *J. Combin. Theory* **2** (1967), 27–34.

11. DeGroot, M. H., Unbiased sequential estimation for binomial populations, *Ann. Math. Statist.* **30** (1959), 80–101.

12. Dongre, N. M., Family trees, *Sankhyā Ser. A*, **33** (1971), 217–220.

13. Enns, E. G., The trivariate distribution of the maximum length, the number of customers served, and the duration of the busy period for the $M/G/1$ queueing system, *J. Appl. Probab.* **6**, (1968), 154–161.

14. Feller, W., *An Introduction to Probability Theory and its Applications*, Vol. 1, 2nd ed. Wiley, New York, 1957.

15. Gabriel, K. R., The distribution of the number of successes in a sequence of dependent trials (abstract), *Ann. Math. Statist.* **30** (1959), 612.

16. Girshick, M. A., Mosteller, F., and Savage, L. J., Unbiased estimates for certain binomial sampling plans with applications, *Ann. Math. Statist.* **17** (1946), 13–23.

17. Good, I. J., Generalizations to several variables of Lagrange's expansion with applications to stochastic processes, *Proc. Cambridge Philos. Soc.* **56** (1960), 367–380.

18. Good, I. J., The generalization of Lagrange's expansion and the enumeration of trees, *Proc. Cambridge Philos. Soc.* **61** (1965), 499–517.

19. Gupta, H. C., and Seth, A., Random walk in the presence of absorbing barriers, *Proc. Nat. Inst. Sci. India A*, **32** (1966), 472–480.

20. Gupta, R. C., Distribution of the sum of independent decapitated generalized negative binomial variables, *Sankhyā Ser. B* **36** (1974), 67–69.

21. Harary, F., Prins, G., and Tutte, W. T., The number of plane trees, *Indag. Math.* **26** (1964), 319–327.

22. Harris, T. E., First passage and recurrence distributions, *Trans. Amer. Math. Soc.* **73** (1952), 471–486.

23. Jain, G. C., On power series distributions associated with Lagrange expansion, *Biometrische Z* **17** (1975), 85–97.

24. Jain, G. C., Some results in a correlated random walk, *Canad. Math. Bull.* **14** (1971), 341–347.

25. Jain, G. C., and Consul, P. C., A generalized negative binomial distribution, *SIAM J. Appl. Math.* **21** (1971), 501–513.

26. Janardan, K. G., Chance mechanism for multivariate hypergeometric models, *Sankhyā Ser. A*, **35** (1973), 465–478.

27. Janardan, K. G., and Patil, G. P., A unified approach for a class of multivariate hypergeometric models, *Sankhyā Ser. A* **34** (1972), 363–376.

28. Kagan, A. M., Linnik, Yu. V., and Rao, C. R., *Characterization Problems in Mathematical Statistics*. Wiley, New York, 1973.

29. Klarner, D. A., A correspondence between two sets of trees, *Indag. Math.* **31** (1969), 292–296.

30. Klarner, D. A., Correspondence between plane trees and binary sequences, *J. Combin. Theory* **9** (1970), 401–411.

31. Kryscio, R. J., and Severo, N. C., Computational and estimation procedures in multidimensional right-shift processes and some applications, *Adv. in Appl. Probab.* **7** (1975), 349–383.

32. Mohanty, S. G., Some convolutions with multinomial coefficients and related probability distributions, *SIAM Rev.* **8** (1966), 501–509.

33. Mohanty, S. G., On a generalized two-coin tossing problem, *Biometrische Z.*, **8** (1966), 266–272.

34. Mohanty, S. G., An r-coin tossing game and the associated partition of generalized Fibonacci numbers, *Sankhyā Ser. A* **29** (1967), 207–214.

35. Mohanty, S. G., The distribution of the maximum queue length, the number of customers served and the duration of the busy period for the queueing system M/M/1 involving batches, *INFOR Canad. J. Oper. Res. Inform. Process.* **9** (1972), 161–166.

36. Mohanty, S. G., On queues involving batches, *J. Appl. Probab.* **9** (1972), 430–435.
37. Mohanty, S. G., and Jain, J. L., On two types of queueing process involving batches, *Canad. Oper. Res. Soc. J.* **8** (1970), 38–43.
38. Moon, J. W., *Counting Labelled Trees*, Canadian Mathematical Monographs, No. 1. Canadian Mathematical Congress, 1970.
39. Mullin, R. C., On counting rooted triangular maps, *Canad. J. Math.* **17** (1965), 373–382.
40. Mullin, R. C. The enumeration of Hamiltonian polygons in triangular maps, *Pacific J. Math.* **16** (1966), 139–145.
41. Mullin, R. C., On the average number of trees in certain maps, *Canad. J. Math.* **18** (1966), 33–41.
42. Narayana, T. V., A partial order and its applications to probability theory, *Sankhyā*, **21** (1959), 91–98.
43. Narayana, T. V., and Mohanty, S. G., Some properties of compositions and their applications to probability and statistics II, *Biometrische Z.* **5** (1963), 8–18.
44. Narayana, T. V., and Sathe, Y. S., Minimum variance unbiased estimation in coin tossing problems, *Sankhyā Ser. A* **23** (1961), 183–186.
45. Narayana, T. V., Chorneyko, I., and Sathe, Y. S., Sufficient partitions for a class of coin tossing problems, *Biometrische Z.* **2** (1960), 269–275.
46. Narayana, T. V., Mohanty, S. G., and Ladouceur, J. C., A combinatorial problem and its applications to probability theory—II, *J. Indian Soc. Agric. Statist.* **12** (1960), 182–189.
47. Rényi, A., On the enumeration of search codes, *Acta Math. Acad. Sci. Hungar.* (1970), 27–33.
48. Riordan, J., *An Introduction to Combinatorial Analysis*, Wiley, New York, 1958.
49. Riordan, J., Ballots and trees, *J. Combin. Theory* **6** (1969), 408–411.
50. Riordan, J., Ballots and plane trees, *J. Combin. Theory Ser. A* **11** (1971), 85–88.
51. Sathe, Y. S., Use of series expansions in estimation problems for distributions involving more than one parameter (abstract), *Ann. Math. Statist.* **30** (1959), 613.
52. Shenton, L. R., and Consul, P. C., On bivariate Lagrange and Borel–Tanner distributions and their use in queueing theory, *Sankhyā Ser. A* **35** (1973), 229–236.
53. Sibuya, M., Yoshimura, I., and Shimizu, R., Negative multinomial distribution, *Ann. Inst. Statist. Math.* **16** (1964), 409–426.
54. Takács, L., A generalization of the ballot problem and its application in the theory of queues, *J. Amer. Statist. Assoc.* **57** (1962), 327–337.
55. Takács, L., A single-server queue with recurrent input and exponentially distributed service times, *Oper. Res.* **10** (1962), 395–399.
56. Takács, L., A combinatorial method in the theory of queues, *SIAM J. Appl. Math.* **10** (1962), 691–694.
57. Takács, L., Combinatorial methods in the theory of queues, *Rev. Internat. Statist. Inst.* **32** (1964), 207–219.
58. Takács, L., *Combinatorial Methods in the Theory of Stochastic Processes*, Wiley, New York, 1967.
59. Tutte, W. T., The number of planted plane trees with a given partition, *Amer. Math. Monthly* **71** (1964), 272–277.
60. Tutte, W. T., On the enumeration of two-coloured, rooted and weighted plane trees, *Aequationes Math.* **4** (1970), 143–156.
61. Zacks, S., *The Theory of Statistical Inference*, Wiley, New York, 1971.

6 Convolution Identities and Inverse Relations

1. Introduction

The present chapter stands apart from the rest since its content involves neither counting formulas nor their applications. However, it still dwells on lattice path combinatorics and similar topics, such as combinatorial identities arising out of convolutions of paths [for example, (4.2)–(4.4)] and inverse relations related to these identities.

Riordan's book [10] will serve as an excellent introduction to inverse relations and their usefulness in combinatorics. The interest in the material covered here stems from a series of papers by Gould [3, 4, 5, 6] on inverse relations, a good part of which however is contained in [10]. The content of this chapter is a continuation and generalization of Gould's work, and is essentially related to lattice path combinatorics. Thus it might serve as a supplement to [10], which offers a great wealth of material on the subject, though our approach is different and is based on paths.

The convolution identities (4.2) and (4.3) and the corresponding generating functions (4.52) and (4.53) are of great help in deriving many probability distributions. The proofs of these identities are simple from the path point of view when a, b and μ are positive integers. But because

165

of the application, we obtain the identities in the general case as well. It is known that the convolution identities can give rise to orthogonal relations which in turn yield inverse relations. Therefore it is fitting to look into these simultaneously in the context of lattice paths, which is the subject of study in the subsequent sections.

2. Convolution Identities

The simplest but most well-known and widely used combinatorial identity is

$$\sum_{k=0}^{n} \binom{a}{k}\binom{b}{n-k} = \binom{a+b}{n}, \quad \text{for any} \quad a, b \quad \text{and integer} \quad n \geq 0,$$

(6.1)

which is called Vandermonde's convolution or Euler's summation formula (see [10, p. 8]). When a and b are positive integers, (6.1) simply means that the number of paths from $(0, 0)$ to $(a + b - n, n)$ is equal to the sum of the number of such paths that pass through $(a - k, k)$, $k = 0, 1, \ldots, n$. This elementary interpretation in terms of paths results in other interesting identities, viz., (4.2)–(4.4).

However, (6.1) is true in general and we may therefore like to ask the same question for other identities too. One way of establishing (6.1) is to use the method of the generating function, and the same approach leads us to the generating functions (4.52) and (4.53).

LEMMA 1

$$\sum_{k=0}^{\infty} \frac{a}{a + \mu k}\binom{a + \mu k}{k} x^k = y^a,$$

(6.2)

$$\sum_{k=0}^{\infty} \binom{a + \mu k}{k} x^k = \frac{y^{a+1}}{y - \mu(y - 1)},$$

(6.3)

where $y^\mu x - y + 1 = 0$ and $|x| < |(\mu - 1)^{\mu - 1}/\mu^\mu|$.

Proof

Set

$$A_n(a, \mu) = \frac{a}{a + \mu n}\binom{a + \mu n}{n},$$

which we call the A-coefficients. Consider $[a/(a + \mu z)]\binom{a + \mu z}{n}$ as a polynomial of degree $n - 1$ in z. Its nth difference must be zero. Thus we get

$$\sum_{k=0}^{n}(-1)^k\binom{n}{k}\frac{a}{a + \mu k}\binom{a + \mu k}{n} = 0, \qquad n \geq 1,$$

which is equivalent to

$$A_n(a, \mu) = \sum_{k=0}^{n-1}(-1)^{n+k+1}\binom{n}{k}\frac{a}{a + \mu k}\binom{a + \mu k}{n}, \qquad n \geq 1. \quad (6.4)$$

$$\sum_{k=0}^{\infty} A_k(a, \mu)w^k$$

$$= 1 + \sum_{k=1}^{\infty} w^k \sum_{j=0}^{k-1}(-1)^{j+k+1}\binom{k}{j}\frac{a}{a + \mu j}\binom{a + \mu j}{k} \qquad \text{(by (6.4))}$$

$$= 1 + \sum_{j=0}^{\infty}(-1)^j\frac{a}{a + \mu j}\sum_{k=j}^{\infty}(-1)^k\binom{k + 1}{j}\binom{a + \mu j}{k + 1}w^{k+1}$$

$$= 1 + \sum_{j=0}^{\infty}(-1)^j A_j(a, \mu)\sum_{k=j}^{\infty}(-1)^k\binom{a + \mu j - j}{k + 1 - j}w^{k+1}$$

$$= 1 - \sum_{j=0}^{\infty} A_j(a, \mu)w^j\sum_{k=1}^{\infty}(-1)^k\binom{a + \mu j - j}{k}w^k.$$

Therefore,

$$\sum_{j=0}^{\infty} A_j(a, \mu)w^j\sum_{k=0}^{\infty}(-1)^k\binom{a + \mu j - j}{k}w^k = 1,$$

which becomes

$$\sum_{j=0}^{\infty} A_j(a, \mu)w^j(1 - w)^{a + j(\mu - 1)} = 1. \quad (6.5)$$

Putting $1 - w = 1/y$ and examining the convergence, we complete the proof of (6.2).

In the region of absolute convergence, we differentiate (6.2) with respect to x and get

$$\sum_{j=1}^{\infty}\binom{a + \mu j - 1}{j - 1}x^{j-1} = \frac{y^{a+\mu}}{y - \mu(y - 1)}.$$

By changing $j - 1$ to k and $a + \mu - 1$ to a, we obtain (6.3), and this completes the proof.

The generating function (6.2) can indeed be established by the Lagrange inverse series expansion (see (1.20) and (1.21), also Skalsky [12]), where the expansion of y^a is obtained in terms of the power series of x from the relation

$$xy^\mu - y + 1 = 0, \tag{6.6}$$

which is known in advance. On the other hand, we have followed Gould [3] and shown in the proof that the series that is the generating function of $A_n(a, \mu)$ is equal to y^a, and (6.6) has resulted as an essential byproduct of our derivation. It is important to remember that in deriving (6.2) and (6.3) no restriction is put on a and μ.

The identity $y^{a+c} = y^a \cdot y^c$ and (6.2) give rise to the following generalization of Vandermonde's convolution identity:

$$\sum_{k=0}^{n} A_k(a, \mu)A_{n-k}(c, \mu) = A_n(a + c, \mu). \tag{6.7}$$

Similarly, a combination of (6.2) and (6.3) leads to another identity:

$$\sum_{k=0}^{n} A_k(a, \mu) \frac{c + \mu(n - k)}{c} A_{n-k}(c, \mu) = \frac{a + c + \mu n}{a + c} A_n(a + c, \mu). \tag{6.8}$$

Another useful identity [3], which can be derived from (6.2), is

$$\sum_{k=0}^{n} (b + dk)A_k(a, \mu)A_{n-k}(c, \mu) = (b(a + c) + nad)A_n(a + c, \mu). \tag{6.9}$$

Remarkably, (6.7) and (6.8) are special cases of (6.9).

By using the generating function and (6.2), it is possible to prove [5]

$$\sum_{k=0}^{n} (-1)^k A_k(a, (1 - \mu)b)A_{n-k}(c + bk - k, \mu b) = A_n(c - a, \mu b). \tag{6.10}$$

When $b = 1$, (6.10) simplifies to (6.7). On the other hand, by putting $\mu = 0$, (6.10) leads to (6.8).

Interestingly, if we consider

$$B_n(a, \mu) = \frac{a}{a + \mu n} \frac{(a + \mu n)^n}{n!}$$

which are called the B-coefficients, it can be checked [4] that (6.7) and (6.8) are true when A-coefficients are replaced by B-coefficients. Identities with A-coefficients are called Vandermonde-type identities while those with B-coefficients are called Abel-type identities. For a detailed discussion see [3, 4]. However, the identity equivalent to (6.10) in terms of B-coefficients is [5]

$$\sum_{k=0}^{n} (-1)^k B_k(a, (1-\mu)b) B_{n-k}(c + bk, \mu b) = B_n(c - a, \mu b). \quad (6.11)$$

It was already seen (Chapter 1, Exercise 10) that

$$\frac{a}{a + \sum_{i=1}^{r} \mu_i n_i} \binom{a + \sum_{i=1}^{r} \mu_i n_i}{n_1, \ldots, n_r} \quad (6.12)$$

has a lattice path interpretation when a and the μ_is are positive integers. Denoting (6.12) by $A(a; \boldsymbol{\mu}, \mathbf{n}; r)$, where no restriction is placed on a and the μ_is, we can get identities [8] similar to (6.7) and (6.8) as

$$\sum_{k_r=0}^{n_r} \cdots \sum_{k_1=0}^{n_1} A(a; \boldsymbol{\mu}, \mathbf{k}; r) A(c; \boldsymbol{\mu}, \mathbf{n} - \mathbf{k}; r) = A(a + c; \boldsymbol{\mu}, \mathbf{n}; r) \quad (6.13)$$

and

$$\sum_{k_r=0}^{n_r} \cdots \sum_{k_1=0}^{n_1} A(a; \boldsymbol{\mu}, \mathbf{k}; r) \frac{c + \sum_{i=1}^{r} \mu_i(n_i - k_i)}{c} A(c; \boldsymbol{\mu}, \mathbf{n} - \mathbf{k}; r)$$

$$= \frac{a + c + \sum_{i=1}^{r} \mu_i n_i}{a + c} A(a + c; \boldsymbol{\mu}, \mathbf{n}; r), \quad (6.14)$$

when we proceed in a way similar to that used before. In fact the more general identity of type (6.10) can be established for $A(a; \boldsymbol{\mu}, \mathbf{n}; r)$. Also, identities involving

$$B(a; \boldsymbol{\mu}, \mathbf{n}; r) = \frac{a}{a + \sum_{i=1}^{r} \mu_i n_i} \frac{(a + \sum_{i=1}^{r} \mu_i n_i)^{\sum_{i=1}^{r} n_i}}{\prod_{i=1}^{r} n_i!},$$

which generalizes the B-coefficient, can be written [8].

Next we turn to (4.4), which covers both (6.7) and (6.8) as special cases, and ask the same question; that is, whether or not the convolution identity is true irrespective of its lattice path interpretation. In order to get an answer we develop the following setting [9].

Consider a sequence $\{f_n(x)\}$ of functions such that

$$f_{-r}(x) = 0, \qquad r > 0;$$

$$f_0(x) = 1; \quad (6.15)$$

$$f_n(0) = \delta_0^n = \begin{cases} 1 & \text{for } n = 0, \\ 0 & \text{for } n \neq 0. \end{cases}$$

(δ_m^n is the Kronecker delta.) The generating function

$$\phi(s, x) = \sum_{n=0}^{\infty} f_n(x)s^n$$

of $\{f_n(x)\}$ is said to be additive with respect to x if

$$\phi(s, x + \alpha) = \phi(s, x)\phi(s, \alpha).$$

Observe that $\phi(s, 0) = 1$ because of (6.15).

Define the g-coefficients recursively by the relation

$$g(b_i, \ldots, b_n) = f_1(b_n)g(b_i, \ldots, b_{n-1}) - f_2(b_{n-1})g(b_i, \ldots, b_{n-2})$$
$$+ \cdots + (-1)^{n-i-1}f_{n-i}(b_{i+1})g(b_i) + (-1)^{n-i}f_{n-i+1}(b_i)$$

$$(6.16)$$

for $i = 1, \ldots, n, n = 1, 2, \ldots$, where the b_is are real numbers (compare (6.16) with (1.30)). Evidently

$$g(b_i, \ldots, b_n) = \|c_{ij}\|_{(n-i+1) \times (n-i+1)}, \qquad (6.17)$$

where $c_{ij} = f_{j-i+1}(b_{n-j+1})$, from which we note that it also satisfies another recurrence relation

$$g(b_i, \ldots, b_n) = f_1(b_i)g(b_{i+1}, \ldots, b_n) - f_2(b_i)g(b_{i+2}, \ldots, b_n)$$
$$+ \cdots + (-1)^{n-i-1}f_{n-i}(b_i)g(b_n) + (-1)^{n-i}f_{n-i+1}(b_i)$$

$$(6.18)$$

for $i = 1, \ldots, n, n = 1, 2, \ldots$.

Denote by $M_m(\mathbf{b})$ and $G_m(\mathbf{b}; \mathbf{a})$ the triangular matrices

$$\begin{pmatrix} f_0(b_1) & (-1)f_1(b_1) & (-1)^2f_2(b_1) & \cdots & (-1)^mf_m(b_1) \\ & f_0(b_2) & (-1)f_1(b_2) & \cdots & (-1)^{m-1}f_{m-1}(b_2) \\ & & f_0(b_3) & \cdots & (-1)^{m-2}f_{m-2}(b_3) \\ & 0 & & \ddots & \vdots \\ & & & & f_0(b_{m+1}) \end{pmatrix}$$

and

$$\begin{pmatrix} 1 & g(a_1 - b_1) & g(a_1 - b_1, a_2 - b_1) & \cdots & g(a_1 - b_1, \ldots, a_m - b_1) \\ & 1 & g(a_2 - b_2) & \cdots & g(a_2 - b_2, \ldots, a_m - b_2) \\ & & 1 & \cdots & g(a_3 - b_3, \ldots, a_m - b_3) \\ & 0 & & \ddots & \vdots \\ & & & & 1 \end{pmatrix}$$

respectively, for any $m \geq 0$. In the following we see that the recurrence relations (6.16) and (6.18) can be alternatively expressed, which is needed in proving Theorem 1. In [1, Section 4] the matrix approach is also suggested for the special case of paths.

LEMMA 2

$M_m(\mathbf{b})$ and $G_m(\mathbf{0}; \mathbf{b})$ are inverses of each other for any b_i and any $m \geq 0$.

We write $M(\mathbf{b})$ and $G(\mathbf{b}; \mathbf{a})$ for $M_\infty(\mathbf{b})$ and $G_\infty(\mathbf{b}; \mathbf{a})$, respectively. In Theorem 1 we give a generalized version of Vandermonde's convolution, which corresponds to (4.4).

THEOREM 1

A necessary and sufficient condition for the matrix identity

$$G(\mathbf{0}; \mathbf{b})G(\mathbf{b}; \mathbf{a}) = G(\mathbf{0}; \mathbf{a}) \qquad (6.19)$$

to hold for any a_i and b_i is that $\phi(s, x)$ be additive with respect to x.

Proof

Let S be the column matrix with transpose $S' = (1, s, s^2, \ldots)$. The identity

$$G(\mathbf{0}; \mathbf{b})M(\mathbf{b})S = S,$$

which is due to Lemma 2, can be written as

$$G(\mathbf{0}; \mathbf{b}) \begin{pmatrix} \phi(-s, b_1) \\ s\phi(-s, b_2) \\ s^2\phi(-s, b_3) \\ \vdots \end{pmatrix} = S, \qquad (6.20)$$

by noting that $M(\mathbf{b})S$ is equal to the second matrix on the left-hand side. Premultiplying $M(\mathbf{b})$ and postmultiplying $M(\mathbf{a})S$ on both sides of (6.19), we get

$$G(\mathbf{b}; \mathbf{a}) \begin{pmatrix} \phi(-s, a_1) \\ s\phi(-s, a_2) \\ s^2\phi(-s, a_3) \\ \vdots \end{pmatrix} = M(\mathbf{b})S \qquad (6.21)$$

by Lemma 2. When $b_i = \beta$ for all i, $G(\mathbf{b}; \mathbf{a})$ becomes $G(\mathbf{0}; \mathbf{c})$ with $c_i = a_i - \beta$ and $M(\mathbf{b})$ becomes $\phi(-s, \beta)$. Comparing (6.21), where $b_i = \beta$, with (6.20), where $b_i = c_i$, we must have

$$\phi(-s, a_i) = \phi(-s, \beta)\phi(-s, a_i - \beta)$$

for all i. This establishes the necessity of the condition.

Observe that (6.18) is equivalent to

$$s^i\phi(-s, b_{i+1}) + \sum_{n=i+1}^{\infty} s^n g(b_{i+1}, \ldots, b_n)\phi(-s, b_{n+1}) = s^i$$

for all i, which after the replacement of b_r by $a_r - b_{i+1}$, $r = i + 1$, $i + 2, \ldots$, becomes

$$s^i\phi(-s, a_{i+1} - b_{i+1})$$

$$+ \sum_{n=i+1}^{\infty} s^n g(a_{i+1} - b_{i+1}, \ldots, a_n - b_{i+1})\phi(-s, a_{n+1} - b_{i+1}) = s^i$$

$$(6.22)$$

for all i. Assuming ϕ to be additive and multiplying both sides of (6.22) by $\phi(-s, b_{i+1})$, we have

$$s^i\phi(-s, a_{i+1}) + \sum_{n=i+1}^{\infty} s^n g(a_{i+1} - b_{i+1}, \ldots, a_n - b_{i+1})\phi(-s, a_{n+1})$$

$$= s^i\phi(-s, b_{i+1}) \qquad (6.23)$$

for all i. In matrix notation (6.23) is identical with (6.21), which is also equal to

$$G(\mathbf{b}; \mathbf{a})M(\mathbf{a}) = M(\mathbf{b}). \qquad (6.24)$$

Finally, when we premultiply $G(\mathbf{0}; \mathbf{b})$ and postmultiply $G(\mathbf{0}; \mathbf{a})$ on both sides of (6.24), we get (6.19) by applying Lemma 2. Thus the sufficiency part is justified, which completes the proof.

We consider three special cases, viz., when $f_r(x) = x$, $\binom{x}{r}$, and $x^r/r!$, the generating functions of which are $x/(1 - s)$, $(1 + s)^x$, and e^{sx}, respectively. Letting the corresponding g-coefficients be g_1, g_2, and g_3, we see from the theorem that g_2 and g_3 satisfy the convolution identity (6.19), whereas g_1 does not. $\binom{x}{r}$ is well known to be a binomial coefficient and $x^r/r!$ is called an Abel coefficient. For special values of b_i, g_2 and g_3 are equal to the A-coefficient and the B-coefficient, respectively.

The matrix identity (6.19) is equivalent to the relations

$$g(a_i - b_i, \ldots, a_n - b_i)$$

$$+ \sum_{j=i}^{n-1} g(a_{j+1} - b_{j+1}, \ldots, a_n - b_{j+1}) g(b_i, \ldots, b_j) + g(b_i, \ldots, b_n)$$

$$= g(a_i, \ldots, a_n) \quad \text{for} \quad i = 0, 1, \ldots, n, \quad n = 2, 3, \ldots, \quad (6.25)$$

and

$$g(a_n - b_n) + g(b_n) = g(a_n) \quad \text{for} \quad i = n, \quad n = 1, 2, \ldots,$$

which is a generalization of Abel-type or Vandermonde-type convolution for g-coefficients.

Since the additive property of $\phi(s, x)$ is equivalent to

$$\sum_{k=0}^{n} f_k(x) f_{n-k}(\alpha) = f_n(x + \alpha), \quad (6.26)$$

which is a convolution identity for $\{f_n(x)\}$, it is therefore remarkable to observe from the theorem that the convolution identity on the fs is sufficient for obtaining the same type of identity on the g-coefficients which are defined by (6.16), (6.17), or (6.18).

3. Orthogonal Relations and Inversion Formulas

A special case of the Vandermonde convolution (6.1) is a simple orthogonal relation

$$\sum_{k=0}^{n} \binom{a}{k} \binom{-a}{n-k} = \binom{0}{n} = \delta_0^n, \quad n \geq 0, \quad (6.27)$$

from which the following set of relations can be written:

$$x_n = \sum_{k=n}^{m} \binom{a}{k-n} y_k \quad \text{if and only if} \quad y_n = \sum_{k=n}^{m} \binom{-a}{k-n} x_k \quad (6.28)$$

for any $m \geq n$. The pair of relations (6.28) is called inverse relations. To check (6.28) we verify that

$$\sum_{k=n}^{m} \binom{a}{k-n} \sum_{j=k}^{m} \binom{-a}{j-k} x_j = \sum_{k=0}^{m-n} \binom{a}{k} \sum_{j=0}^{m-n-k} \binom{-a}{j} x_{j+k+n}$$

$$= \sum_{r=0}^{m-n} x_{r+n} \sum_{k=0}^{r} \binom{a}{k} \binom{-a}{r-k} = x_n$$

because of (6.27), and similarly,

$$\sum_{k=n}^{m} \binom{-a}{k-n} \sum_{j=k}^{m} \binom{a}{j-k} y_j = y_n$$

because of (6.27).

The above example reveals the interplay between orthogonal relations and inverse relations. To get a more comprehensive picture of it, the reader is advised to refer to [10]. Quite clearly, we may be able to generate an orthogonal relation from a convolution identity, which on the other hand forms the basis of an inversion formula. Other illustrations [5] follow. It we put $c = a$ in (6.10) and (6.11), we get orthogonal relations, the inversion formulas corresponding to which are

$$x_n = \sum_{k=n}^{m} (-1)^{k-n} A_{k-n}(a + nb - n, (1 - \mu)b) y_k,$$

$$y_n = \sum_{k=n}^{m} A_{k-n}(a + nb - n, \mu b) x_k; \tag{6.29}$$

and

$$x_n = \sum_{k=n}^{m} (-1)^{k-n} B_{k-n}(a + nb, (1 - \mu)b) y_k,$$

$$y_n = \sum_{k=n}^{m} B_{k-n}(a + nb, \mu b) x_k. \tag{6.30}$$

In general, given two sets $\{A(t, n)\}$ and $\{B(t, n)\}$ of numbers satisfying $A(t, n) = B(t, n) = 0$ when $n < t$ or $n < 0$ or $t < 0$, we say that the number sets are orthogonal is

$$\sum_{k=t}^{n} A(1, k) B(k, n) = \delta_t^n \tag{6.31}$$

for all t and n. In matrix notation (6.31) takes the form

$$A_m B_m = I_m \qquad \text{for any} \quad m \geq 0, \tag{6.32}$$

where I_m is the $(m + 1)$th order identity matrix and A_m, B_m are the $(m + 1)$th order triangular matrices with the (i, j)th element equal to $A(i - 1, j - 1)$, $B(i - 1, j - 1)$, respectively for $i \leq j$. Since (6.32) implies

$$B_m A_m = I_m \qquad \text{for any} \quad m \geq 0, \tag{6.33}$$

we have

$$\sum_{k=t}^{n} B(t, k)A(k, n) = \delta_t^n$$

for all t and n, which means the orthogonality is a symmetric relation. Letting $X'_m = (x_0, x_1, \ldots, x_m)$, $Y'_m = (y_0, y_1, \ldots, y_m)$ for any $m \geq 0$, we immediately conclude that

$$X_m = A_m Y_m \quad \text{if and only if} \quad Y_m = B_m X_m, \tag{6.34}$$

which are inverse relations. Also, since

$$A'_m B'_m = I_m = B'_m A'_m, \tag{6.35}$$

we get another set of inverse relations, viz.,

$$X_m = A'_m Y_m \quad \text{and} \quad Y_m = B'_m X_m. \tag{6.36}$$

In some situations (6.36) may not give any new set of relations; for example, (6.28) will not change. The second set of inversion formulas corresponding to (6.29) is

$$x_n = \sum_{k=0}^{n} (-1)^{n-k} A_{n-k}(a + bk - k, (1 - \mu)b)y_k,$$

$$y_n = \sum_{k=0}^{n} A_{n-k}(a + bk - k, \mu b)x_k. \tag{6.37}$$

Orthogonal relations may not always arise through convolution identities. It is noteworthy to state the most frequently mentioned example (see [10, p. 4]) that

$$\sum_{s=m}^{n} (-1)^{s+m} \binom{s}{m} \binom{n}{s} = \delta_m^n, \tag{6.38}$$

the inverse relations of which give rise essentially to the well-known method of inclusion and exclusion in counting [13]. Lemma 2 also provides such an example, where

$$M_m(\mathbf{b})G_m(0; \mathbf{b}) = I_m = G_m(0; \mathbf{b})M_m(\mathbf{b}). \tag{6.39}$$

Obviously, we can write inversion formulas induced by (6.39). Besides the special cases cited in the previous section, it would be of interest to study the inverse relations when $f_r(x) = A_r(x, d)$ or $B_r(x, d)$.

In our examples orthogonal relations are obtained from (6.1), (6.10), and (6.11) by making the right-hand side equal to zero for $n > 0$. Now let us look for an orthogonal relation from the convolution identity (6.25) of g-coefficients. Observing from (6.17) that when $b_i = 0$,

$g(b_i, \ldots, b_n) = 0$ and letting $a_i = 0$ in (6.25), we get the identities in matrix notation as

$$C_m(\mathbf{b})G_m(\mathbf{b}; \mathbf{a}) + D_m(\mathbf{b}; \mathbf{a}) = I_m, \qquad m \geq 0, \qquad (6.40)$$

where $C_m(\mathbf{b})$ is the same as $G_m(\mathbf{0}, \mathbf{b})$ except that its diagonal elements are zeros, and $D_m(\mathbf{b}; \mathbf{a})$ is a triangular matrix given by

$$\begin{pmatrix} 1 & g(-b_1) & g(-b_1, a_2 - b_1) & g(-b_1, a_2 - b_1, a_3 - b_1) & \cdots & g(-b_1, a_2 - b_1, \ldots, a_m - b_1) \\ & 1 & g(-b_2) & g(-b_2, a_3 - b_2) & \cdots & g(-b_2, a_3 - b_2, \ldots, a_m - b_2) \\ & & 1 & g(-b_3) & \cdots & g(-b_3, a_4 - b_3, \ldots, a_m - b_3) \\ & & & 1 & \cdots & g(-b_4, a_5 - b_4, \ldots, a_m - b_4) \\ & 0 & & & \ddots & \\ & & & & & 1 \end{pmatrix}.$$

Notice that (6.40) is not of the form of (6.32) and therefore is not the matrix of orthogonal relations in the earlier sense. Alternatively, (6.40) can be expressed as

$$(C_m(\mathbf{b}), I_m)\begin{pmatrix} G_m(\mathbf{b}; \mathbf{a}) \\ D_m(\mathbf{b}; \mathbf{a}) \end{pmatrix} = I_m, \qquad m \geq 0. \qquad (6.41)$$

Thus we trivially know that if

$$X_{2m+1} = \begin{pmatrix} G_m(\mathbf{b}; \mathbf{a}) \\ D_m(\mathbf{b}; \mathbf{a}) \end{pmatrix} Y_m, \qquad (6.42)$$

then

$$Y_m = (C_m(\mathbf{b}), I_m)X_{2m+1}, \qquad (6.43)$$

which cannot be considered as the set of inverse relations in the usual way. Also notice that, if we assume $a_j = 0$ for all j in (6.25), we get

$$G_m(\mathbf{0}; \mathbf{b})G_m(\mathbf{b}; \mathbf{0}) = G_m(\mathbf{0}; \mathbf{0}) = I_m, \qquad m \geq 0. \qquad (6.44)$$

It is left as an exercise to check that $G_m(\mathbf{b}; \mathbf{0}) = M_m(\mathbf{b})$ and that hence (6.44) is reduced to (6.39).

The situation (6.39) motivates us to generalize the definition of orthogonality. A system $\{A_{1m}\}, \ldots, \{A_{km}\}$ of $(m + 1)$th order triangular matrices is called orthogonal to another system $\{B_{1m}\}, \ldots, \{B_{km}\}$ of $(m + 1)$th order triangular matrices if

$$\sum_{i=1}^{k} A_{im}B_{im} = I_m, \qquad m \geq 0. \qquad (6.45)$$

The generalization of inversion as in the preceding paragraph is obviously one-sided.

4. Concluding Remarks

We may examine the convolution identities and inversion formulas in the light of Rota's work [11] on the Möbius inversion formula. The incidence algebra of a locally finite partially ordered set P is defined as follows. Consider the set of all real-valued functions of two variables $f(x, y) = 0$, defined for $x \in P$ and $y \in P$, such that $f(x, y) = 0$ if $x \nleq y$. The sum of two such functions f and g and multiplication by scalars are defined as usual. The product $h = fg$ is given by

$$h(x, y) = \sum_{x \leq z \leq y} f(x, z)g(z, y).$$

It can be seen that the incidence algebra has an identity element which is δ_x^y (i.e., $\delta_x^y = 0$ if $x \neq y$ and $\delta_x^x = 1$) and that the inverse of the zeta function $\zeta(x, y)$, which is defined to be $\zeta(x, y) = 1$ if $x \leq y$ and $\zeta(x, y) = 0$ otherwise, is the Möbius function $\mu(x, y)$, which is defined inductively as $\mu(x, x) = 1$ and $\mu(x, y) = -\sum_{x \leq z < y} \mu(x, z)$ for $x < y$. In [11] the Möbius inversion formula (Corollary 1 of [11]) states the following:

Let $f(x)$ be a real-valued function defined for $x \in P$ such that $f(x) = 0$ unless $x \leq x^* \in P$.

If

$$g(x) = \sum_{x \leq y \leq x} f(y),$$

then

$$f(x) = \sum_{x \leq y \leq x} \mu(x, y)g(y).$$

Let P be a locally finite completely ordered set. Clearly, because of identity (6.25), we may say that an incidence algebra is defined on the g-coefficients. We have seen that in (6.17) when

$$c_{ij} = \binom{a_j + 1}{j - i + 1},$$

the g-coefficients satisfy the convolution identity. In its place, consider g-coefficients for the case

$$c_{ij} = \binom{a_j + r}{r + j - i}, \qquad r > 1,$$

as motivated by (2.16). The answer to the question of whether or not an incidence algebra can be established on the new coefficients is negative since the convolution identity is not true. This is noted from the example for $n = 1, r = 2$ that in general

$$\binom{a_1 + 2}{2} \neq \binom{b_1 + 2}{2} + \binom{a_1 - b_1 + 2}{2}.$$

In the particular case of the completely ordered set P (say, P is the set of natural numbers), we observe that

$$\mu(x, k) = \begin{cases} 1 & \text{if } k = x, \\ -1 & \text{if } k = x + 1, \\ 0 & \text{otherwise.} \end{cases}$$

Thus the inversion formula reduces to the following trivial form:

$$g(x) = \sum_{y=x}^{x^*} f(y) \quad \text{if and only if} \quad f(x) = g(x) - g(x + 1). \quad (6.46)$$

In the incidence algebra there might be pairs of functions other than ζ and μ such that one is the inverse of the other. Similarly to the Möbius inversion formula, one should be able to generate other inversion formulas by utilizing these pairs. Unfortunately, for a general P no pair other than ζ and μ is determined in an obvious manner. However, when P is completely ordered, each such pair satisfies the orthogonal relation. In Section 3 we cited several cases of the orthogonal relation, from which we readily obtained inversion formulas that are different from the trivial one, viz., (6.46).

Exercises

1. Setting $i = 1$ and $f_r(x) = \binom{x}{r}$ in (6.16), $N_j = g(b_1, \ldots, b_j)$ and $N_0 = 1$, prove

$$\sum_{j=0}^{\infty} N_j t^j (1 - t)^{b_j + 1} = 1,$$

and show that (6.2) and (6.3) follow from the above relation when we substitute $b_{j+1} = (\mu - 1)j + a$ [2].

2. By proper specialization of g-coefficients, we can get (6.9)–(6.11). Prove these independently.

3. Prove (6.13) and (6.14).

4. Write the identities analogous to (6.10) and (6.11) in terms of $A(a; \mu, \mathbf{n}; r)$ and $B(a; \mu; \mathbf{n}; r)$ and derive the corresponding inverse series relations.

5. Prove the following identities:

(a) $$\sum_{R_2}\sum_{R_1}\prod_{i=1}^{k}\frac{a_i}{a_i+\mu_1 x_i+\mu_2 y_i}\binom{a_i+\mu_1 x_i+\mu_2 y_i}{x_i,\, y_i}$$
$$=\frac{\sum_{i=1}^{k}a_i}{\sum_{i=1}^{k}a_i+\mu_1 n_1+\mu_2 n_2}\binom{\sum_{i=1}^{k}a_i+\mu_1 n_1+\mu_2 n_2}{n_1,\, n_2},$$

where
$$R_1=\left\{(x_1,\ldots,x_k)\colon \sum_{i=1}^{k}x_i=n_1,\, x_i\geq 0 \quad \text{for all}\ i\right\}$$

and
$$R_2=\left\{(y_1,\ldots,y_k)\colon \sum_{i=1}^{k}y_i=n_2,\, y_i\geq 0 \quad \text{for all}\ i\right\}.$$

(b) $$\sum_{k=1}^{n}\sum_{C_k}(-1)^k\prod_{i=1}^{k}\frac{c}{c+\mu r_i}\binom{c+\mu r_i}{r_i}=\frac{-c}{-c+\mu n}\binom{-c+\mu n}{n},$$

where
$$C_k=\left\{(r_1,\ldots,r_k)\colon \sum_{i=1}^{k}r_i=n, r_i\geq 1\right\}.$$

[Hint: Set $a=-c$ in (6.7) and find an expression for $(-c/(-c+\mu n))\binom{-c+\mu n}{n}$ which is used repeatedly for various values of n.]

6. Prove
$$\sum_{k_r=0}^{n_r}\cdots\sum_{k_1=0}^{n_1}\frac{a+\sum_{i=1}^{r}\mu_i k_i}{a}B(a;\mu,\mathbf{k};r)$$
$$\times\frac{c-\sum_{i=1}^{r}\mu_i k_i}{c-\sum_{i=1}^{r}\mu_i n_i}B\!\left(c-\sum_{i=1}^{r}\mu_i n_i,\mu,\mathbf{n}-\mathbf{k};r\right)$$
$$=\sum_{k_r=0}^{n_r}\cdots\sum_{k_1=0}^{n_1}\frac{\sum_{i=1}^{r}k_i}{\prod_{i=1}^{r}k_i!}B(a+c;\mathbf{0},\mathbf{n}-\mathbf{k};r)\prod_{i=1}^{r}\mu_i^{k_i}.$$

(This is a generalization of Jensen's convolution formula [4, 7]).

7. A generalized version of the method of inclusion and exclusion. Let S be an arbitrary set and F a field of subsets of S. Let V be a finite and finitely additive set function defined on F (i.e., $V(A_1 + A_2) = V(A_1) + V(A_2)$ if $A_1 \cap A_2 = \varnothing$, $A_1 \in F$, $A_2 \in F$). Given $A_1, \ldots, A_n$ belonging to F, we denote by H_k the set of elements that belong to exactly k sets among $A_1, \ldots, A_n$. By using (6.38) prove that [13]

$$V(H_k) = \sum_{r=k}^{n} (-1)^{r-k} \binom{r}{k} U_r, \qquad k = 0, 1, \ldots, n,$$

where $U_0 = V(S)$ and

$$U_r = \sum_{i \le i_1 < i_2 < \cdots < i_r \le n} V(A_{i_1} \cap A_{i_2} \cap \cdots \cap A_{i_r}), \qquad r = 1, \ldots, n.$$

8. Show that the g-coefficients with $f_n(x) = A_n(x, \mu)$ or $f_n(x) = B_n(x, \mu)$ satisfy the convolution identity (6.25).

9. Prove the following inversion formulas:

(a) $\displaystyle x_n = \sum_{r=n+1}^{m} g(b_{n+1}, \ldots, b_r) y_r + y_n$

if and only if

$$y_n = \sum_{r=n}^{m} (-1)^{r-n} f_{r-n}(b_{n+1}) x_r$$

for any $n \le m, m \ge 0$;

(b) $\displaystyle x_n = \sum_{s=0}^{n-1} g(b_{s+1}, \ldots, b_n) y_s + y_n$

if and only if

$$y_n = \sum_{s=0}^{n} (-1)^{n-s} f_{n-s}(b_{s+1}) x_s$$

for $n \ge 0$.

10. What are the inverse relations induced by (6.45)?

References

1. Carlitz, L., Roselle, D. P., Scoville, R. A., Some remarks on ballot-type sequences of positive integers, *J. Combin. Theory* **11** (1971), 258–271.

2. Göbel, F., Some remarks on ballot problems, *Math. Centrum Amsterdam*, Rep. S 321a, 1964.

3. Gould, H. W., Some generalizations of Vandermonde's convolution, *Amer. Math. Monthly* **63** (1956), 84–91.
4. Gould, H. W., Generalization of a theorem of Jensen concerning convolutions, *Duke Math. J.* **27** (1960), 71–76.
5. Gould, H. W., A new convolution formula and some new orthogonal relations for inversion of series, *Duke Math. J.* **29** (1962), 393–404.
6. Gould, H. W., The construction of orthogonal and quasi-orthogonal number sets, *Amer. Math. Monthly* **72** (1965), 591–602.
7. Mohanty, S. G., Some convolutions with multinomial coefficients and related probability distributions, *SIAM Rev.* **8** (1966), 501–509.
8. Mohanty, S. G., and Handa, B. R., Extensions of Vandermonde-type convolutions with several summations and their applications—I, *Canad. Math. Bull.* **12** (1969), 45–62.
9. Mohanty, S. G., and Handa, B. R., A generalized Vandermonde-type convolution and associated inverse relations, *Proc. Cambridge Philos. Soc.* **68** (1970), 459–474.
10. Riordan, J., *Combinatorial Identities*. Wiley, New York, 1968.
11. Rota, G. C., On the foundation of combinatorial theory I. Theory of Möbius functions, *Z. Wahrsch. Verw. Gebiete* **2** (1964), 340–368.
12. Skalsky, M., A note on a convolution-type combinatorial identity, *Amer. Math. Monthly* **74** (1967), 836–838.
13. Takács, L., On the method of inclusion and exclusion, *J. Amer. Statist. Assoc.* **62** (1967), 102–113.

Index

Probability and Mathematical Statistics

A Series of Monographs and Textbooks

Editors   **Z. W. Birnbaum**   **E. Lukacs**

University of Washington   *Bowling Green State University*

Seattle, Washington   *Bowling Green, Ohio*

E. J. McShane. Stochastic Calculus and Stochastic Models. 1974

Robert B. Ash and Melvin F. Gardner. Topics in Stochastic Processes. 1975

Avner Friedman, Stochastic Differential Equations and Applications, Volume 1, 1975; Volume 2. 1975

Roger Cuppens. Decomposition of Multivariate Probabilities. 1975

Eugene Lukacs. Stochastic Convergence, Second Edition. 1975

H. Dym and H. P. McKean. Gaussian Processes, Function Theory, and the Inverse Spectral Problem. 1976

N. C. Giri. Multivariate Statistical Inference. 1977

Lloyd Fisher and John McDonald. Fixed Effects Analysis of Variance. 1978

Sidney C. Port and Charles J. Stone. Brownian Motion and Classical Potential Theory. 1978

Konrad Jacobs. Measure and Integral. 1978

K. V. Mardia, J. T. Kent, and J. M. Biddy. Multivariate Analysis. 1979

Sri Gopal Mohanty. Lattice Path Counting and Applications. 1979

in preparation

Michel Metivier and J. Pellaumail. Stochastic Integration

Y. L. Tong. Probability Inequalities in Multivariate Distributions